Salim Gasmi
Chihab Slimi
Ikram Djareche Boudiar Nachwa

Addiction and Hallucinogens

Salim Gasmi
Chihab Slimi
Ikram Djareche Boudiar Nachwa

Addiction and Hallucinogens

Impacts and Treatments

ScienciaScripts

Imprint
Any brand names and product names mentioned in this book are subject to trademark, brand or patent protection and are trademarks or registered trademarks of their respective holders. The use of brand names, product names, common names, trade names, product descriptions etc. even without a particular marking in this work is in no way to be construed to mean that such names may be regarded as unrestricted in respect of trademark and brand protection legislation and could thus be used by anyone.

Cover image: www.ingimage.com

This book is a translation from the original published under ISBN 978-620-6-69613-1.

Publisher:
Sciencia Scripts
is a trademark of
Dodo Books Indian Ocean Ltd. and OmniScriptum S.R.L publishing group

120 High Road, East Finchley, London, N2 9ED, United Kingdom
Str. Armeneasca 28/1, office 1, Chisinau MD-2012, Republic of Moldova, Europe
Printed at: see last page
ISBN: 978-620-8-11994-2

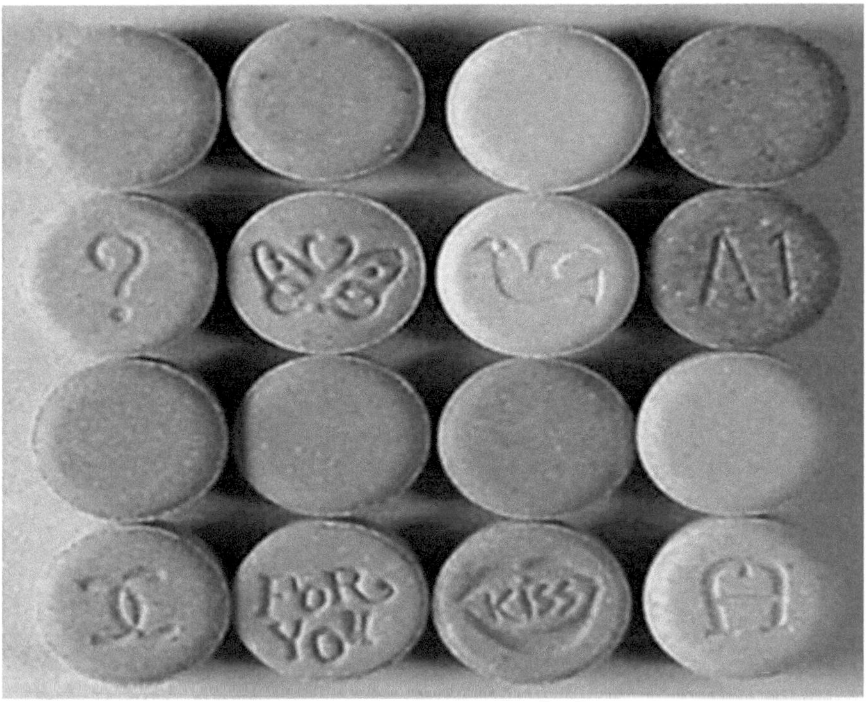

Drug addiction and hallucinogens

Impacts and Treatments

Dr. GASMI Salim

Mr SLIMI Chiheb

Melle . DJARECHE Ikram

Melle . BOUDIAR Nachwa

University of Tébessa

October 2023

Epigraph

"Depression and addiction are the names given to the uncontrollable when it is no longer a question of conquering freedom, but of becoming oneself and taking the initiative to act".

Alain EHRENBERG (1998)

Contents

Summary

The use of psychoactive substances worldwide remains a major public health problem. It has spread in an unusual way in recent times.

The methodology of this book was based on taking information from its source, where a sample of consumers of psychoactive substances (hallucinogens).

The profile of patients is characterised by their young age (48%) of young adults [21-25 years], the predominance of males (82%), against (18%) of females. Pregabalin® was the most commonly used substance (59%), followed by Tramadol® (18%). The distribution of patients according to daily dose shows a high rate of 2 doses per day (35%), with an age of consumption of more than 5 years, central nervous system disorders (33%) and other disorders.

Management is marked by the presence of a relatively good treatment protocol on the ward, despite the absence of excellent conditions for treatment, and drug testing.

Key words: Psychoactive substances, Hallucinogens, Tramadol, Drug addiction, Tébessa.

Introduction

From time immemorial, mankind has been guided by the quest for pleasure. In this quest, man has been supported by numerous psychotropic substances, first extracted from nature, then modified and/or synthesised **(Alain, 2015).**

Addiction and the search for ultimate pleasure distance drug addicts from their day-to-day concerns, neglecting the health and social consequences of their compulsive consumption **(Alain and Grégory, 2019).** Substance abuse is a societal health problem with several repercussions **(Degenhardt, 2018).**

This type of consumption is particularly common among teenagers, during an important developmental period **(Kandel, 1992).** Moreover, failure to control the side effects of the psychoactive substances taken can lead to death (concept of overdose) **(Landie, 2015).**

For some people, the use of psychotropic substances takes the form of escalation, i.e. the tendency to start by using more frequently used substances (tobacco, alcohol, cannabis) and move on to less frequently used substances (Lyrica, amphetamine), some of which require more invasive methods of consumption (nasal, injection) **(Tassel, 2017).** In recent years, drug use in Algeria has expanded remarkably, mainly due to the fragility of the economic and social infrastructure. This is particularly visible in the Tébessa region, although this Wilaya remains isolated and far from drug sources. Unfortunately, Tébessa has been affected by the spread of this scourge.

In addition, corruption has spread, particularly nepotism, within many state structures and institutions, with a lack of effective monitoring and control. In addition, most regions have been marginalised in terms of development projects. All these factors have contributed to a high rate of drug abuse and trafficking, in addition to Algeria's strategic location and its extensive and multiple borders **(Kaddour, 2015).**

To this end, the subject of our study was a survey of the consumption of hallucinogens and their impact on drug addicts (case of the Wilaya of Tébessa), trying to analyse these figures to discover the reasons for the increase in consumption rates, the target group and most types of pills prevalence and percentage of damage to the body.

The first part is devoted to theory (Chapter 1 on drug addiction, Chapter 2 on hallucinogens and Chapter 3 on treatment), and the practical part comprises the various results obtained and their discussion. We end this modest work with a conclusion.

Chapter 01

Drug addiction

1. General

The term addiction comes from the Greek words "Toxikon" (poison by arrow), which gave rise to "toxicum" (poison in Latin), and "mania" (madness). Is a psychological or even physical dependence on one or more chemical substances (psychotropic drugs, synthetic drugs, plants, solvents) with no therapeutic need **(Djeribi, 2018).**

According to the American Society of Addiction Medicine(ASAM), addiction is a chronic primary disease of the brain's reward, motivation and memory areas and associated neural circuits **(Landry *et al.*, 2012).**

The WHO defines drug addiction as a state of physical and/or psychological dependence on a substance, established in a person as a result of its periodic or continuous use.

It prefers the term "drug dependence" to "addiction". According to the International Classification of Diseases and Related Health Problems (ICD 10), dependence can be said to exist when at least three of the following criteria are present over a period of one year and have persisted for at least one month or have occurred repeatedly:

- Powerful desire to consume.
- Difficulty controlling consumption.
- Physical withdrawal syndrome when stopping or reducing consumption.
- Increased tolerance (the dose must be increased to obtain the same effect).
- Gradual divestment of other businesses and consumer bonds.
- Continued use despite harmful consequences **(Audrey *et al.*, 2010).**

Drug addiction is the interaction between the consumer, the psychotropic drug and the socio-cultural context. It is a complex biopsychosocial problem, linked to physical, mental and sociological components **(Amar *et al.*, 2002).**

2. Classification

2.1. Pharmacological

According to **Delay & Deniker (1957)**, there are three groups of psychotropic drugs which may or may not be used as drugs in the addictive sense.

2.1.1. Psycholeptics or sedatives

Decreased physical and mental activity. They include hypnotics (barbiturates), neuroleptics and tranquillisers **(Sandra, 2000).**

2.1.2. Psychoanaleptics or stimulants

Excitement of physical and mental activities. They include vigilance stimulants (amphetamines), mood stimulants (antidepressants or thymoanaleptics) and psychotonics (caffeine, theine) **(Audrey *et al.*, 2010).**

2.1.3. Psychodysleptic or disruptive disorders

Distortion of perceptions and changes in mental functions. They include hallucinogens (LSD), narcotics and intoxicants **(Sandra, 2000).**

2.2. Legal

Three categories of drugs are distinguished by whether or not they are legal.

2.2.1. Legal substances sold under state control (tobacco/alcohol).

2.2.2. Legal *prescription* **drugs** *diverted from their* therapeutic *use:* tranquillisers, barbiturates, amphetamines, etc. **(Audrey *et al.*, 2010)**

2.2.3. Illicit substances: known as narcotics, the list of which is drawn up by an interministerial commission: cannabis, heroin, opiate derivatives, cocaine and psychological stimulants **(Sandra, 2000).**

3. Poly-drug use

Polydrug use is a concept with many different definitions, depending on the study consulted. For some, polydrug use has a temporal perspective, i.e. the simultaneous use of two or more substances or the use of these substances over a prolonged period, without the use being simultaneous. Other researchers define polydrug use in terms of the simultaneous effects that psychotropic drugs can have on an individual, which amounts to a study of the different types of effects of the simultaneous use of psychotropic drugs **(Amar *et al.*, 2002)**. For example, whether the combination of certain substances increases or decreases the effect of each drug, or rather whether the combinations create a new effect. On the other hand, another less studied categorisation of poly-drug use relates to the way in which psychotropic drugs are combined and thus poly-drug used **(Ives & Ghelani, 2006).**

	Benzodiazepines (Kietyl)	**Amphetamines (Ecstasy)**	**Tramadol**	**Ketamine**	**Pregabalin (Lyrica)**	**LSD**
Heroin	Increase the effects of the high or prolong the effect of heroin. Risk of coma and breathing difficulties.	These may be signs of an overdose that could lead to heart problems, coma or even death.	Causes breathing difficulties, and an overdose can lead to death.	Loss of balance, loss of consciousness and, above all, respiratory depression.	Depression Respiratory, Increase the risk of death.	Increased risk of losing touch with reality, unpredictable effect.
Cocaine	Increases the risk of overdose.	The risks of hyperthermia, dehydration and heart problems.	Sudden death, high blood pressure, heart palpitations, respiratory failure, damage to brain tissue.	Coke may increase the heart rate, ketamine may slow it down or increase it, it all depends.	Heart attacks. Dehydrat Brain swelling.	Risk of cardiac arrest.
Cannabis		Anhedonia (inability to feel positive emotions), dysphoria (mood disorders) or depression.	The risk of heart rhythm disorders.	Breathing difficulties.	Respiratory depression Cardiac arrest Coma Deaths	Delirium. Disorganised behaviour. Paranoia. Depression. Backtracking.

Table 01: The risks of mixing drugs (poly-drug use) **(Cirriez, 2015; Dufau, 2020; Matthieu *et al.*, 2022).**

4. Stages of substance use disorders

People who become dependent on substances such as alcohol or drugs generally go through four phases. The four main stages are

4.1 Experiments

Substance abuse begins with experimentation or voluntary consumption. In adolescents and young adults, the substance is usually bought and tested with friends, often in response to peer pressure. Older people may start self-medicating in response to problems in their personal lives, such as separation, divorce, bereavement or job loss (**Neuraxpharm, 2021**).

4.2 Regular or risky consumption

This stage is characterised by frequent use. It does not necessarily occur every day, but the substance may be used in a predictable pattern (for example, every weekend) or in a predictable situation (for example, when bored, lonely, anxious or upset). It starts to become risky when it creates emotional, physical or social problems **(Neuraxpharm, 2021).**

4.3. Dependency

In the addiction stage, the person continues to use the substance regularly despite the physical, social and emotional damage it causes. Here are some of the main signs of dependence:

- Not taking responsibility.
- Drinking or using drugs repeatedly and in dangerous situations.
- Developing tolerance, which means taking more drugs to achieve the same effect.
- Experience withdrawal symptoms **(Sandra, 2000; Neuraxpharm, 2021).**

There are 2 types of addiction:

- **Psychic dependence**: This is a condition in which the drug produces a feeling of satisfaction and a psychic impulse that requires periodic or continuous administration of the drug to produce pleasure or avoid a depressive state. Some authors call it addiction **(Sandra, 2000).**
- **Physical dependence**: This is a state of adaptation which manifests itself in intense disorders when the drug is discontinued. The problems associated with the abstinence syndrome (withdrawal) are psychological or physical in nature, producing the opposite effects to those initially developed by the drug.

(Ciciero & Ellis, 2015).

4.4 Tolerance

Tolerance means that, over time, you have to take larger and larger doses of a drug to obtain the same effects. Over time and with repeated exposure, the body is better able to break down the drug and eliminate it from the bloodstream. It is also less receptive to the presence of the drug. In fact, many long-term users who have developed tolerance are unable to feel the pleasurable effect or sense of well-being of their first take **(Sandra, 2000).**

Drug overdose describes the administration of a drug or other substance in excess of the amount recommended or usually consumed **(Landie, 2015).**

A tolerance threshold that has disappeared or been reduced, e.g. following medical detoxification, poses a risk of overdose. If the same amount of drug is consumed as before detoxification (i.e. while tolerance is reduced), the body is unable to break down and eliminate the drug as quickly as before, hence the risk of overdose **(Reis *et al.,* 2015).**

5. Causes of drug addiction

It's always difficult to know why someone takes drugs in the first place. The reasons for experimenting with a drug are not the same as the reasons for continuing to use it, or even becoming dependent on it:

5.1. Genetic factors

Some people may be genetically more vulnerable to the addictive properties of drugs. Studies have shown that the risk of developing a substance abuse disorder is higher in people who have a close relative with such a disorder **(Marilyn *et al.,* 2010).**

5.2. For fun and relaxation

When a person takes a drug, they expect to experience pleasurable sensations and benefit from it. These recreational uses of drugs or alcohol are usually occasional **(Laurent, 2014).**

5.3. To carry out different experiments

It is the search for sensations and risk-taking that motivates drug use. For example, an adolescent may take a drug simply to experience a new sensation. Their aim is to go beyond what they usually feel **(Ray, 2017 ; Laurent, 2014).**

5.4. Dealing with problems

The aim may be to escape from reality. Drugs are used to numb thoughts, feelings, emotions and suffering. Such use can lead to dependence and a deterioration in the user's situation **(Laurent, 2014).**

5.5. To improve performance

Performance, fitness, energy, coping with the pressure of exams and work are all reasons given for using drugs **(Ray, 2014).**

6. Risk factors for addiction (drug dependence)

There are three components to addiction: the individual, the product and the environment.

6.1. Individual factors

An individual's age, gender, brain maturity, personality and mood play a significant role in their individual risk of addiction **(Laurent, 2017)**.

Early initiation and male gender are specific vulnerabilities. For example, starting to use psychotropic substances in early adolescence increases by a factor of ten the risk of becoming dependent on psychotropic substances in adulthood, compared with starting later, around the age of 20. People who are anxious, introverted or depressive are at greater risk of addiction **(Pajot *et al.*, 2015; Benyamina, 2017).**

From a neurobiological point of view, the level of activity of the neurotransmitters that govern our functioning and behaviour can vary from one individual to another, making some people vulnerable to the risk of addiction **(Benyamina, 2017).**

Disruptions to the dopaminergic (involved in the reward circuit) or serotonin (mood) systems, in particular, are associated with such vulnerability **(Laurent, 2014).**

6.2. Products/practices with variable addictive potential

As far as the product is concerned, addiction can take hold more or less quickly: after one or a few takes (crack, cocaine...), more gradually, or even slowly (psychotropic medication...) **(Benyamina, 2017).**

6.3. Environmental factors

Finally, the influence of the environment (stress, social and friendship context, presence of psychological disorders, etc.) is also a determining factor **(Laurent, 2014).**

7. Symptoms of substance use disorders :

There are a number of signs and symptoms of substance use disorders that are common to most types of addiction. These include:

7.1 Psychological changes

- Sudden mood swings.
- Appearance of a feeling of anxiety, fear or paranoia for no particular reason.
- Concentration problems.
- Low self-esteem.
- Aggravation of existing mental disorders such as anxiety or depression **(Neuraxpharm, 2021).**

7.2. Physical changes

- Sudden weight loss or gain.
- Unusual skin colour.
- Bad breath.
- General deterioration in physical appearance **(Ray, 2017).**

7.3. Behavioural and social changes

- Lack of motivation to do anything, apathy.
- Changes in eating or sleeping habits.
- Loss of interest in hobbies and activities.
- Withdrawal of friends and family.
- Neglecting responsibilities at home or at work **(Neuraxpharm, 2021).**

8. Neurobiology of addiction

8.1. Reward circuit

The reward circuit describes a network of connections that link two groups of neurons, one located in the ventral tegmental area and the other in the nucleus accumbens. The latter is a paired structure **(Pajot *et al.*, 2015)**. There is one in the left hemisphere and another in the right hemisphere. The reward circuit serves to reinforce some of our behaviours: it is involved, for example, in eating or sexual behaviour **(Ray, 2017).**

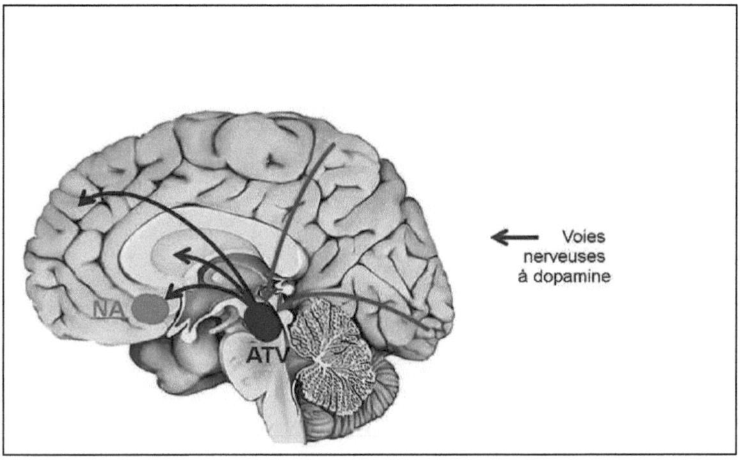

Figure 01: Reward circuit **[web site 01].**

8.1.1. Neurons

The synapse is a junctional area through which the chemical message (neurotransmitter) passes from one neuron to another. It is a particularly important structure, where the neuronal signal can be reinforced (via an increase in neurotransmitter release or in the number of postsynaptic receptors) or decreased (via an increase in the number of presynaptic receptors) **(Pajot *et al.*, 2015).**

8.1.2. Information circuit between two neurons

Our brain secretes nearly a hundred neurotransmitters, such as dopamine and serotonin. The area of communication or connection between two neurons is called a synapse. The neurotransmitter secreted by the first neuron crosses the space between the two neurons and reaches the second neuron where it binds to specific receptors **(Mildeca, 2021).**

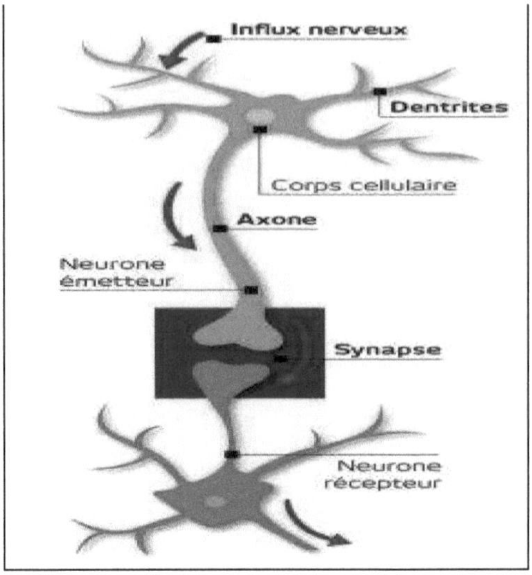

Figure 02: Information circuit between two neurons.

8.2. The reward circuit and addiction

Addiction to a substance develops after it has been used several times. The first time it is used, our olfactory and gustatory cells discover it, and it is only after the second use that our cells' memory kicks in and the substance is recognised. In the brain, the reward circuit is triggered, with certain neurotransmitters such as dopamine, adrenaline, serotonin and acetylcholine playing a key role in this circuit. When addiction occurs, the substance's active ingredient binds to its receptor (a natural process), which releases the neurotransmitters **(Dever, 2015)**. Excessive use of the substance will unbalance this circuit and the patient will want to use more and more of the substance in order, without really knowing it, to release more and more neurotransmitters. When they stop taking the substance, withdrawal symptoms will appear (sweating, trembling, etc.). This imbalance will make you want to take the drug again, so that you don't get the same cravings again when you stop taking it. There is no longer any pleasure in taking the drug, just an escape from the craving: this is known as negative reinforcement. The reward circuit is therefore at the heart of our addiction pathophysiology **(Devaux, 2022).**

Chapter 02:

Hallucinogens

1. General

Hallucinogens are chemical substances that, at low doses, cause changes in perception, thought or mood, while maintaining alertness, attention, memory and orientation. They produce auditory, visual and tactile distortions and hallucinations - a dream-like experience - in waking subjects **(John *et al.*, 2007).**

A hallucinogen is any substance that causes changes in perception to the point of hallucination. It may therefore be a synthetic substance (synthetic hallucinogen) or a plant or mushroom (natural hallucinogen). What they all have in common is that they induce an altered state of consciousness by acting on the central nervous system. An overdose of hallucinogens can be fatal **(Botrel, 2021).**

2. History

Historically, some of these substances have had ancestral ritual uses, some of which have survived to the present day, notably through shamanism and certain cults (Ayahuasca, for example). Their ritual uses are varied. cult, curative, ritual, trance. It was the emergence of the great monotheistic religions that was responsible for the disappearance of these uses, which they designated as embodying evil **(Chambon, 2009)**.

Despite their ancient origins, modern Western society did not take a real interest in these substances until the very early twentieth century, especially after the discovery of LSD and the cultural revolution that followed. These substances were then tested mainly for therapeutic purposes, particularly during psychotherapy. Their growing popularity led to them being banned, and most of these substances are now illegal. There is currently a revival of scientific and medical research in Western countries (USA, Switzerland, Spain), with numerous experimental studies having demonstrated the possible use of certain hallucinogens for medical indications **(Chambon, 2009).**

3. Therapeutic use of hallucinogens

Between 1950 and the mid-1960s, more than a thousand scientific articles were published in clinical psychiatry journals in the USA and Europe, describing the treatment of almost 40,000 patients with hallucinogenic substances. On reading the international literature, the various clinical indications selected for the use of hallucinogens are as follows **(Dever, 2015)**:

- Treatment of post-traumatic mental disorders.

- Treatment of pain and depression in patients at the end of life (cancer patients, AIDS sufferers, etc.) **(Dever, 2015)**.
- Treatment of psychosomatic disorders and obsessive neuroses.
- Treatment of depression.
- Treatment of psychoses (the most controversial indication, gradually abandoned from the 1960s onwards).
- Psycholytic treatment in psychotherapy and oniroanalysis **(Sueur *et al.*, 2017)**.

4. Classification (Clinical classification of hallucinogens)

Hallucinogens are psychotropic drugs that are described as "central nervous system disruptors". Hallucinogens are divided into 3 classes:

4.1. Delusional hallucinogens

They induce a state of delirium, characterised by extreme confusion and an inability to control one's actions. They cause a very strong dissociation of consciousness, but without immobility. These substances are toxic, carry a risk of death from overdose, and also include a number of side effects such as dehydration, mydriasis and heart rhythm disorders **(Dever,2015)**.

4.2. Dissociative hallucinogens

At anaesthetic doses, they produce analgesia, amnesia and catalepsy. At subanaesthetic doses, they produce a feeling of detachment from the environment, and a dissociation of body and mind. The soul separates from the body, and the individual has the impression that it is floating above the physical body. Dissociative symptoms include disturbance of memory, identity, distance estimation and perception **(Fleurentin, 2015)**.

Other dissociation experiences include an inability to recognise oneself in the mirror while maintaining awareness that the image in the mirror is one's own. They include ketamine **(Fleurentin, 2015)**.

4.3. Psychedelic hallucinogens

A psychedelic substance is a psychoactive substance that causes alterations in cognition and perception. The psychedelic effect is characteristic of this class. They include LSD **(Fleurentin, 2015)**.

5. Some examples of hallucinogens

5.1. Cannabis

5.1.1. Definition

Cannabis (cannabis sativa) or Indian hemp is a small tree native to Asia. The main active chemical ingredient in cannabis is THC (delta-9-tetrahydrocannabinol). Of the 400 or so chemical ingredients found in the cannabis plant, THC is the one that most affects the brain. It is a psychotropic chemical that brings a feeling of euphoria (high) to people who consume cannabis **(Dever, 2015).**

Cannabis in different forms:

- **Marijuana:** leaves, stems and flowering tops, simply dried.
- **Hashish:** compressed resin obtained from the plant.
- **Oil: a** preparation with a higher concentration of active ingredients, obtained by distilling the resin.

And can be consumed in different ways: Inhalation (smoking or vaping) and ingestion (eating or drinking) are two common ways **(Dever, 2015).**

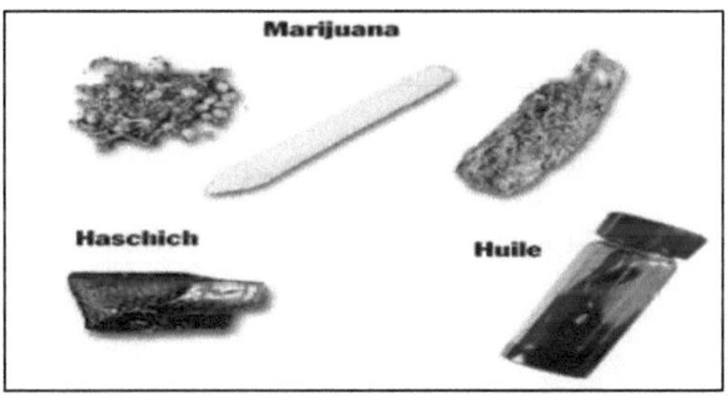

Figure 03: Forms of cannabis use **(Clermont *et al.*, 2014).**

5.1.2. Chemical structure

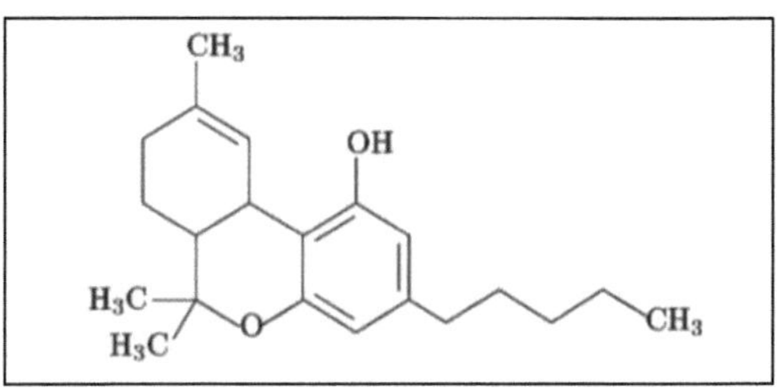

Figure 04: Chemical structure of cannabis **(Paczesny, 2014).**

5.1.3. Pharmacology

5.1.3.1. Mechanism of action

Action on central (CB1) or peripheral (CB2) stereospecific receptors **(Vincent, 2017).**

5.1.3.2. Kinetics

- **Absorption:** Rapid by pulmonary route (2 to 3 minutes), slower by oral route (2 to 4 hours).

- **Distribution:** Approximately 90% of THC in the blood is distributed to plasma, where it is highly bound to plasma proteins (97-99%), mainly lipoproteins. THC rapidly penetrates vascularised tissues such as the liver, heart, fat, kidneys, placenta and pituitary gland. Its plasma concentration then decreases rapidly **(Fabritius, 2014).**

- **Metabolism:** THC is mainly metabolised in the liver.

- **Elimination:** THC is excreted, mainly in the form of inactive metabolites, in the faeces and to a lesser extent in the urine.

    - First elimination half-life of 8 days.

    - Second late elimination half-life of 2 to 3 months **(Fabritius, 2014).**

5.1.4 Effects of cannabis

1. **Risks associated with occasional consumption**

- Movement disorders, dizziness, muscle relaxation, sometimes drowsiness.

- Hallucinations may occur (if large amounts of THC are taken quickly, particularly with oil).

- Dry mouth, red eyes, mydriasis.

- Stimulation of hunger after consumption **(Dever, 2015).**

2. **Risks associated with chronic consumption**
    - **When** smoked, cannabis poses the same **respiratory** risks as tobacco (irritation, probable increased risk of respiratory cancers), especially when mixed with tobacco. Cannabinoid syndrome Chronic abdominal pain, cyclical nausea and vomiting (every 2-3 months), often in the morning **(Dever, 2015).**
    - **Cardiovascular** changes in heart rate (tachycardia) and blood pressure **(Dever, 2015).**
    - **Periodontal (gum disease) and oral diseases** gingivitis (inflammation of the gums), periodontitis (inflammation of the tissues surrounding the tooth, leading to loosening of the teeth and subsequent tooth loss), stomatitis (inflammation of the oral cavity), gingival hyperplasia (enlargement of the gums) **(Dever, 2015).**

Cannabis use increases the risk of mental illness, particularly schizophrenia **(Vandendriessche *et al.*, 2010).**

5.2. Cocaine

5.2.1. Definition

Cocaine generally takes the form of a fine, white, crystalline, odourless powder. It is extracted from coca bush leaves. When snorted, it is known as "line coke"; it is also sometimes injected intravenously or smoked, mainly in the form of crack (or free base) **(De Collective, 2006).**

Figure 05: Cocaine powder **(DEA, 2017).**

5.2.2. The form of cocaine

Cocaine exists mainly in 2 forms:

- Cocaine hydrochloride: it comes in the form of colourless crystals, a crystalline powder (snow) or grains (rock), and is taken intravenously or, more frequently, intranasally.
- The other form is crack, the alkaloid in base form obtained by heating cocaine hydrochloride in the presence of bicarbonate of soda and water; it can also be smoked mixed with tobacco **(Burnat & Le Brumant, 1992).**

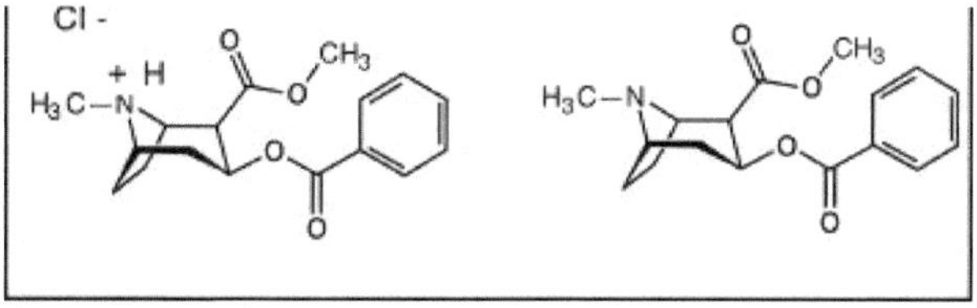

Figure 06: Cocaine and crack cocaine molecules **(Carl *et al.*, 2014)**

5.2.3. Pharmacokinetics of cocaine

- It reaches the brain in 3 minutes when taken nasally.
- Intravenously, the cocaine is completely absorbed and the effect takes 15 seconds.

➢ It goes through the alveoli into the arterial circulation in the form of smoked and inhaled crack and reaches the brain in less than 8 seconds.

➢ Cocaine diffuses into all the body's tissues, easily crossing the blood and brain barriers.

➢ It is eliminated by several processes:

- Mainly by metabolism: metabolites have a high proportion in the urine and can be detected in the blood and urine 3 to 4 days after being taken.
- Unaltered in urine (cocaine).
- Par la bile **(Chaumont, 2018).**

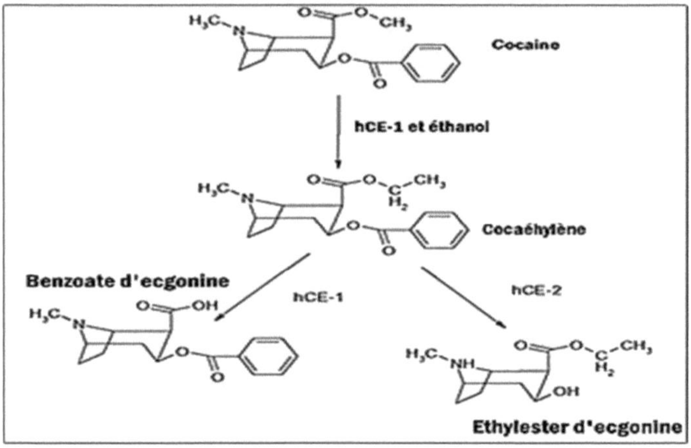

Figure 07: Cocaine metabolism **(Bensakhria, 2015).**

5.2.4. Mode of action

Cocaine blocks the reuptake of several neurotransmitters, mainly dopamine and to a lesser extent serotonin and noradrenaline. It activates dopaminergic neurons in several regions of the brain, particularly in the nucleus acumbens, a structure that plays an important role in regulating emotions. The privileged role of dopaminergic neurons in the reflex activation of the reward circuit has now been clearly established, and is certainly linked to the triggering of craving **(Gothuey, 2014).**

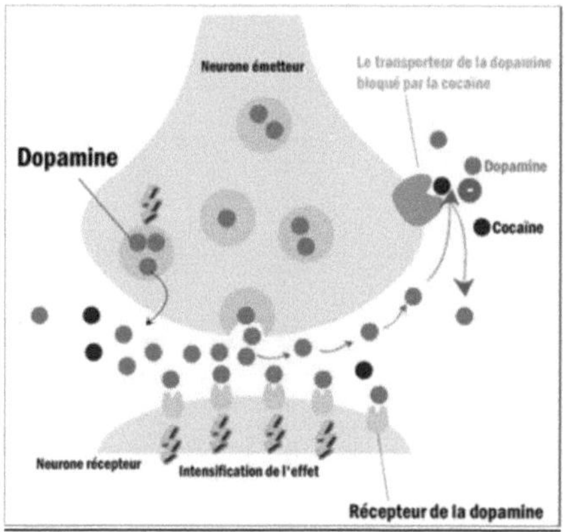

Figure 08: Mechanism of action of cocaine **(Bensakhria,2015).**

5.2.5 Effects

- **Cardiovascular:** acute rhabdomyolysis, necrosis, ischaemia, myocardial infarction.
- **Central Nervous System:** Meningeal haemorrhage, ictus, headache.
- **Respiratory system:** Pulmonary haemorrhage, dyspnoea, pneumopathy.
- **Renal:** Acute renal failure.
- **Hepatic:** Liver infection (HBC, HCV) **(Gothuey, 2014).**

5.3. Ecstasy

5.3.1. Definition

MDMA (methyl-dioxy-methylamphetamine) is a member of the phenethylamine family, like amphetamine. It acts on the dopaminergic systems, giving it stimulant properties. It circulates in the form of tablets known as "ecstasy" **(Gandilhon *et al.,* 2016).**

A drug classified as a narcotic and increasingly used by young people for its stimulant and hallucinogenic effects (**Burnat, 1996**).

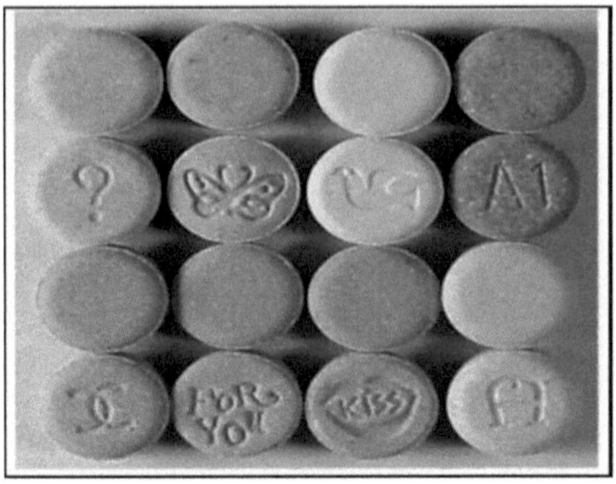

Figure 09: Variety of ecstasy tablets **(Lapostolle *et al.*, 2021).**

2.3.2. Chemical structure

Ecstasy is an amphetamine derivative with the chemical formula C11H5N02 belonging to the phenylethylamine family. Its scientific name is methylenedioxy3,4dimethylphenylethylamine or methylenedioxy-3,4- methamphetamine or MDMA **(Dussaud, 1998).**

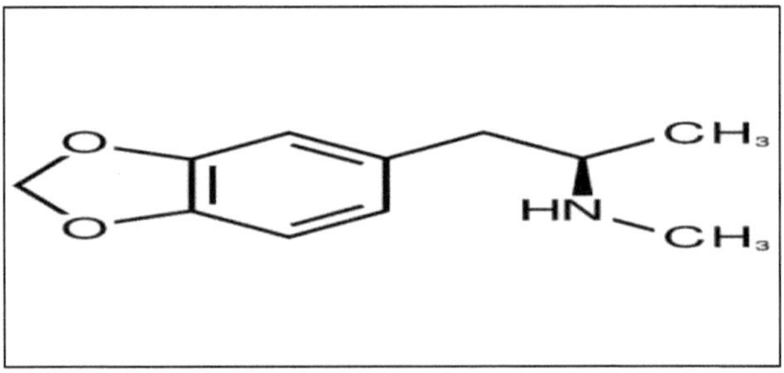

Figure 10: Chemical structure of MDMA **(Dussaud, 1998)**

5.3.3. Pharmacology

5.3.3.1. Mechanism of action

MDMA targets dopamine and serotonin neurons. MDMA is thought to act by blocking the dopamine and serotonin transporters responsible for reuptaking these neurotransmitters. The result is an immediate, significant and transient increase in dopamine and serotonin in the synaptic clefts of the corresponding synapses, and consequent hyperactivation of dopamine and serotonin receptors in the central and peripheral nervous systems **(Constance, 2002).**

5.3.3.2. Kinetics

- Ecstasy is well **absorbed** from the digestive tract. Amphetamines have a low molecular weight and low protein binding, favouring rapid tissue distribution. The volume of distribution is high. Peak plasma levels are reached within one to two hours.

- **The metabolism of** MDMA is now well described. The first and quantitatively most important route involves opening the methylenedioxy ring, followed by methylation or glucuro- or sulphoconjugation. The second route involves dealkylation to the biologically active (methylene dioxyamphetamine) MDA. Deamination and oxidation then produce benzoic acid and derivatives excreted in the form of hippuric acid after conjugation with glycine. Metabolites (including quinones) are produced in this way, and their neurotoxicity has been clearly established. Other metabolites have been identified as having hepatic, renal and cardiac toxicity **(Lapostolle *et al.*, 2021).**

- Amphetamines are mainly excreted in the urine **(Lapostolle *et al.*, 2021)**

5.3.4. Effects

- MDMA causes mydriasis (dilated pupils), bruxism (clenched jaws, muscle tremors), nausea and vomiting, and a rise in blood pressure. Taking MDMA can be particularly hard on the body (especially the heart, liver and kidneys). Increased body temperature can lead to heat stroke and dehydration. An overdose can cause cramps and hallucinations.
- MDMA can cause epilepsy.
- Consumption of MDMA is toxic to the liver, and can be particularly dangerous if you have heart problems, asthma, epilepsy or diabetes.
- Excessive or frequent consumption can lead to alterations in the serotonin system, which can result in cognitive problems or transient or more persistent depressive symptoms **(addictaide ,2022).**

5.4. Heroin

5.4.1. Definition

Heroin is a semi-synthetic opiate. Morphine, the main psychoactive principle in opium, has been modified to form diacetylmorphine, which differs mainly in that it is more fat-soluble, enabling it to cross the blood-brain barrier more quickly. Heroin exists in two forms: a hydrocholrure (heroin-HCL) used for injection because of its solubility in water, and an alkaline derivative (heroin-base) which can be vaporised and therefore consumed in cigarettes. Heroin has a number of characteristics: it is a powerful painkiller and anxiolytic, and has strong antitussive properties **(Lebeau, 2002).**

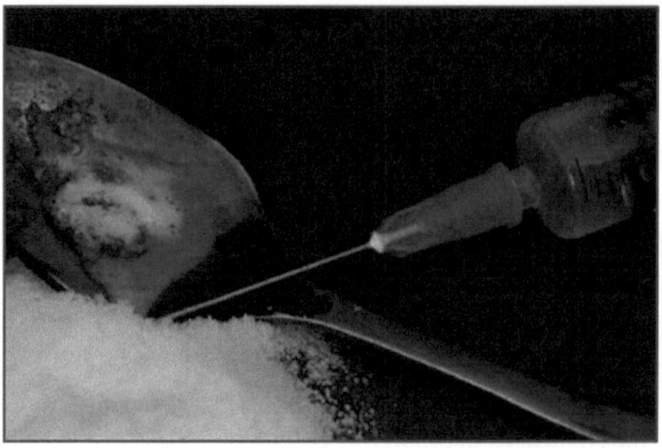

Figure 11: Heroin **(Dominique, 2010).**

5.4.2 Form

There are three forms:

- Basic heroin: is not soluble in water and cannot be injected; it is only smoked.
- Heroin hydrochloride: i.e. adding acid hydrochloride to the basic heroin, for injection.
- In powder form: soluble in water, it is composed of 30-50% pure heroin, aspirin caffeine and heroin powder or (browen-sugar), mix it with lemon juice or vinegar to consume **(Chaumont, 2018)**.

5.4.3 Pharmacokinetics of heroin

Once in the brain, it is rapidly hydrolysed into morphine, which is the form that is actually active at the receptors **(Denis et al., 1998).**

The metabolism of heroin, like that of other natural opiates, is rapidly deacetylated into 6-monoacetylmorphine and then morphine. As a result, heroin has a short plasma half-life of 3 to 10 minutes. There is no trace of heroin in the urine. The end product of heroin catabolism is morphine, whose half-life varies from 1.5 to 4 hours. It undergoes conjugation reactions, making it more water-soluble, as well as oxidation reactions, transforming it into codeine. It is therefore classically found in urine **(Denis et al., 1998).**

5.4.4 Mechanism of action

Heroin has a molecular structure similar to endorphins (dynorphin). It then binds to endorphin receptors and increases the release of dopamine, resulting in a euphoric, calming, pain-relieving effect **(Chaumont, 2018)** .

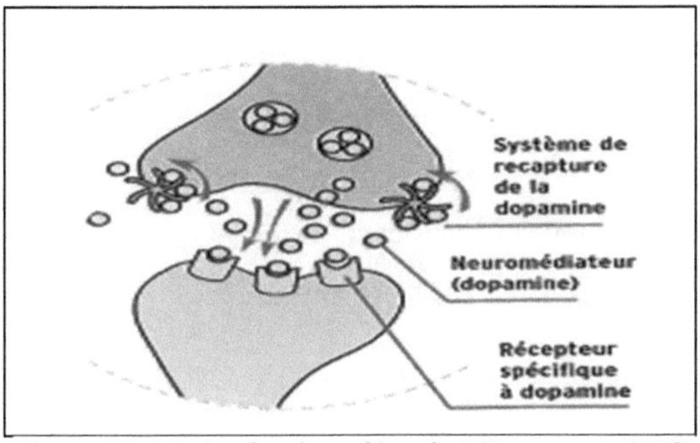

Figure 12: Mode of action of heroin **(Chaumont, 2018)**

5.4.5. Effects of heroin use

- **Immediate effects**

➢ Dysphoria, nausea and vomiting and reduced appetite.

➢ Causes a sensation of warmth in the skin.

➢ Dry mouth and sweating **(Demaret *et al.*, 2013).**

- **Effects according to method of injection**

➢ Intravenously; the injection process damages the venous system

➢ Sharing equipment causes diseases such as HIV, hepatitis C and B, tetanus and tuberculosis **(Demaret *et al.*, 2013).**

- **Long-term effects**

➢ A problem of low bone mass and sometimes osteoporosis.

➢ A deleterious effect on heart cells **(Demaret *et al.*, 2013).**

5.5. Psychotropic drugs

5.5.1. Tramadol

5.5.1.1. Definition

Tramadol is a short-acting opioid1, a synthetic derivative of codeine. It is sometimes described as a "weak" opioid **(Gruffat, 2022).**

Is an analgesic widely used in the management of painful patients **(Devaux,2022).**

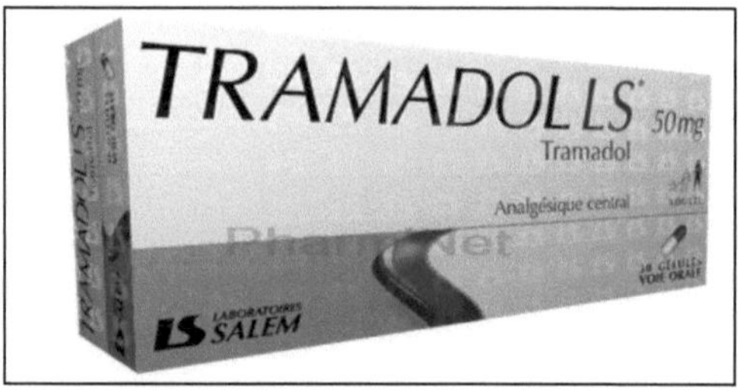

Figure 13: Commercial form of tramadol in Algeria **(diagnostics.labosalem.dz, 2023).**

5.5.1.2. Chemical structure

Tramadol, with the empirical formula C16H25NO2, has a structure similar to codeine and morphine (Figure 14). Its chemical formula is (1RS, 2RS) -2-[(dimethylamino)- methyl] -1-(3-methoxyphenyl)-cyclohexanol **(Cassier, 2014).**

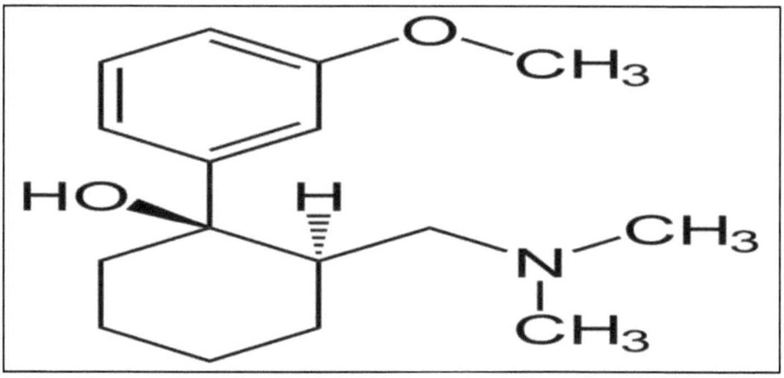

Figure 14: Chemical structure of tramadol **(Cassier, 2014)**

5.5.1.3. Dosage

This medicine containing 100 mg tramadol is indicated primarily for severe pain. For less intense pain, a medicine containing 50 mg tramadol **(Gruffat, 2022)** is recommended.

- Acute pain: The initial dose is 100 mg (1 tablet) followed by 50 to 100 mg every 4-6 hours, without exceeding 400 mg / 24 hours (4 tablets).

- Chronic pain: The initial dose is 50 or 100 mg (1 tablet) followed by 50 to 100 mg (1 tablet) every 4-6 hours, without exceeding 400 mg / 24 hours (4 tablets) **(Gruffat, 2022).**

5.5.1.4. Pharmacological properties

5.5.1.4.1. Pharmacokinetics

It corresponds to the fate of the drug in the body and comprises 4 phases:

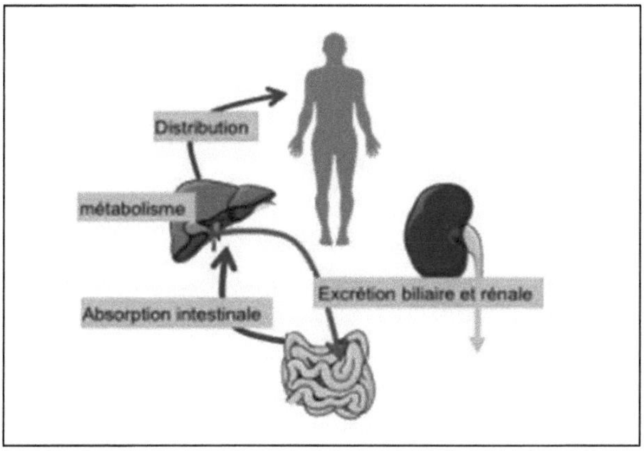

Figure N15: The fate of the drug in 4 stages **(Devaux, 2022).**

Absorption

Tramadol is absorbed orally with a bioavailability of over 90% when taken repeatedly every 6 hours. For a single oral administration, bioavailability is between 70 and 90%. Peak serum levels (Tmax) are reached in approximately 2 hours.

Distribution

Tramadol is poorly bound to plasma proteins (20%) and has a high volume of distribution (3 to 4 litres/kg). In addition, tramadol will pass through the placenta and will be present in very small quantities in breast milk.

Metabolism

In the case of tramadol, 90% of the drug is metabolised mainly by the liver.

Elimination

The elimination half-life of tramadol varies between 5 and 7 hours. Tramadol is 95% eliminated in the kidneys, with the remainder eliminated in the faeces **(Devaux, 2022).**

5.5.1.4.2. Pharmacodynamics

Tramadol is a Tier II central opioid analgesic (acting in the brain and spinal cord). However, it is not a morphine derivative and has a dual mechanism of action, resulting in synergy of action at therapeutic doses:

- Central monoaminergic effect resulting in inhibition of the reuptake of serotonin and noradrenaline. It is this effect that is active on neuropathic pain.

- Opioid effect due to binding to μ-type opioid receptors. It is a partial and non-selective agonist of μ, d and k morphine receptors, with a more intense affinity for μ receptors **(Devaux, 2022).**

5.5.1.5 Inductions

Tramadol is indicated for the management of moderate to severe pain **(Gruffat, 2022).**

5.5.1.6. Counter inductions

Like all medicines, tramadol is not without its drawbacks. These include

- Hypersensitivity to tramadol and opiates.
- Severe respiratory failure.
- Severe liver failure.
- Epilepsy not controlled by treatment.
- Breast-feeding if long-term treatment is necessary.
- Acute intoxication or overdose with CNS depressants and/or alcohol.
- A combination of morphine agonists/antagonists and MAOIs. **(Devaux,2022).**

5.5.1.7 Undesirable effects

The most frequently encountered side effects are those of the opiate class, such as :

- Nausea, vomiting.
- Drowsiness, headaches, dizziness, hypersudation, feeling unwell,
- Dry mouth.
- Constipation if taken for long periods **(Schaad, 2001).**

5.5.2. Lyrica (Pregabalin)

5.5.2.1. Definition

Pregabalin is approved as an adjunctive treatment for partial-onset seizures, peripheral and central neuropathic pain and fibromyalgia **(Spigset & Westin, 2013).**

Pregabalin (Lyrica) has been described as "an analogue of the neurotransmitter γ-aminobutyric acid (GABA) with analgesic, anticonvulsant and anxiolytic effects" **(Tasson *et al.*, 2007).**

BMP misuse and abuse potential, which can lead to substance dependence and intoxication (Figure 16) **(Hagg *et al.*, 2020).**

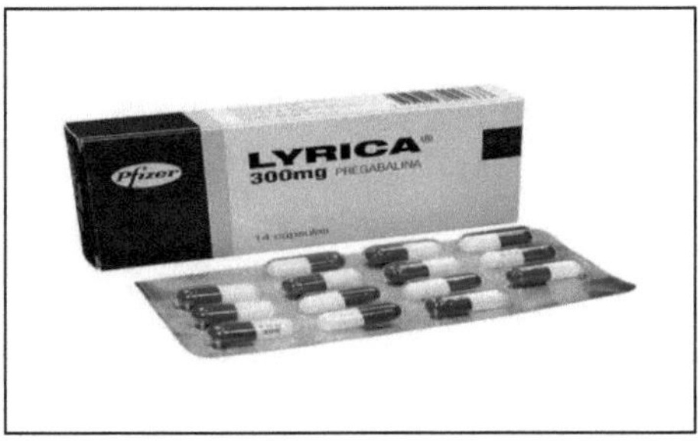

Figure 16: The commercial form of pregabalin in Algeria **(Ammari *et al.*, 2021).**

5.5.2.3. Chemical structure

Pregabalin is the S enantiomer of 3-aminomethyl-5-methylhexanoic acid (figure17) **(Dooley *et al.*, 2002).**

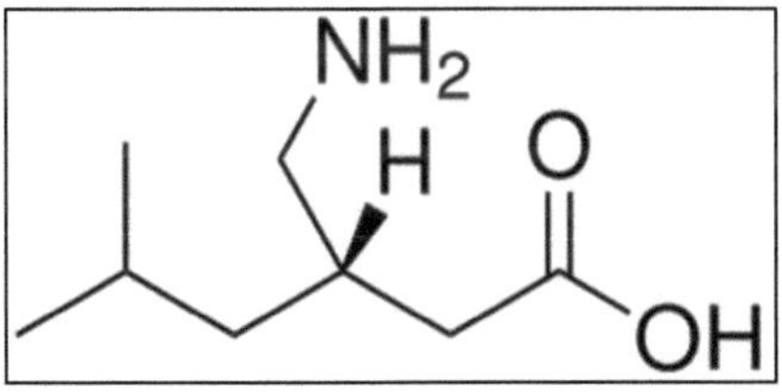

Figure 17: Chemical structure of PGB **(Dooley *et al.*, 2002).**

5.5.2.4. Pharmacokinetics

Absorption

Pregabalin is rapidly absorbed after oral administration, with peak plasma concentrations occurring between 0.7 and 1.3 hours. The oral bioavailability of pregabalin is approximately 90% and is independent of dose and frequency of administration. Food reduces the rate of pregabalin absorption, resulting in lower and delayed peak plasma concentrations, but the extent of drug absorption is unaffected, suggesting that pregabalin can be administered regardless of meals **(Howard *et al.*, 2013) .**

Distribution

-It is fairly well distributed throughout the body, with very little binding to plasma proteins.

-It is capable of crossing the blood-brain and placental barriers and is excreted in breast milk **(Palma and Pancheri, 2007).**

Metabolism

Pregabalin is not metabolised in the liver; it is eliminated from the systemic circulation mainly by renal excretion in unchanged form. Its clearance is directly proportional to creatinine clearance, and is unaffected by age, sex, race, menopause and concomitant administration of other anti-epileptic drugs **(Bock brader *et al.*, 2010).**

Elimination

The elimination half-life of pregabalin is approximately 6 hours and steady-state is reached in 1 to 2 days after repeated administration. Adjusted for oral bioavailability, the plasma clearance of pregabalin is essentially equivalent to renal clearance, indicating that pregabalin undergoes negligible non-renal elimination **(Howard *et al.*, 2013).**

Pregabalin has desirable and predictable pharmacokinetic properties that suggest ease of use. As pregabalin is renally eliminated, renal function affects its pharmacokinetics **(Howard *et al.*, 2013).**

5.5.2.5. Pharmacodynamics

Preclinical research using receptor autoradiography and replacement mutations **(Li *et al.*, 2011).**

Have shown that pregabalin binds selectively to the α2δ subunits of voltage-gated calcium channels, resulting in a decrease in the release of neurotransmitters (norepinephrine, glutamate, substance P, calcitonin gene-related peptide) and a reduction in postsynaptic excitability **(Dooley *et al.*, 2003).**

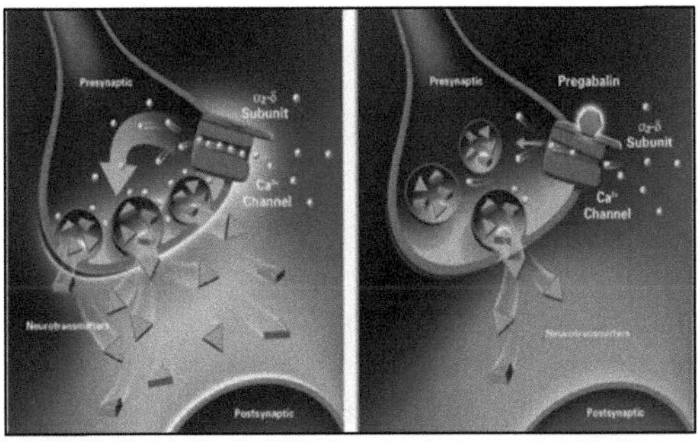

Neuronal hyperexcitability	Neuron modulation with pregabalin

Figure 18: Mechanism of action of pregabalin **(Bian *et al.*, 2006)**.

Potential-gated calcium channels are cell membranes that are permeable to calcium ions, which control the action potential of selective neurons **(Bian *et al.*, 2006)**.

The left-hand side of the image illustrates the moment when neurons enter through voltage-dependent calcium channels and activate the release of neurotransmitters, allowing the vesicles to join the cell membrane of the synaptic cleft. An excess influx of calcium ions occurs. The right-hand side of the image illustrates the strong binding of pregabalin to alpha2-delta subunits **(Shim, 2011)**.

5.5.2.6. Dosage

Table 02: Dosage of pregabalin **(Parke-Davis Div of Pfizer Inc., 2004)**.

Capsules	25mg	50mg	75mg	100mg	150mg	200mg	225mg	300mg
Drinkable solution	20mg/ml							

Dosage varies from 150 to 600 mg a day, in two or three doses. Neuropathic pain, Epilepsy, Generalised anxiety disorder. Treatment with pregabalin may be initiated at a dose of 150 mg per day, administered in two or three doses. Depending on the patient's response and tolerance, the dose may be increased to 300 mg per day after an interval of 3 to 7 days, and may if necessary be increased to the maximum dose of 600 mg per day after a further interval of 7 days **[website 03]**.

5.5.2.7 Indication

- Neuropathic pain. Lyrica is indicated for the treatment of peripheral and central neuropathic pain in adults.

- Epilepsy. Lyrica is indicated in combination in adults for the treatment of partial epileptic seizures with or without secondary generalisation.

- Generalised anxiety disorder. Lyrica is indicated for the treatment of generalised anxiety disorder (GAD) in adults. **Website 04**

5.5.2.8. Contraindication

- LYRICA is contraindicated in patients with known hypersensitivity to pregabalin or any of its components. Reports of hypersensitivity reactions, including cases of angioedema, have been received **(Parke-Davis Div of Pfizer Inc., 2004)**. The occurrence of symptoms of angioedema such as facial, perioral or upper airway swelling requires immediate discontinuation of pregabalin.

- The use of Lyrica during the first trimester of pregnancy may cause major congenital malformations in the unborn child.

- Treatment with pregabalin has been associated with dizziness and drowsiness, which could increase the risk of accidental injuries (falls) in the elderly population. **(Le Figaro Santé., 2004)** .

5.5.2.9. Undesirable effects

The main neurological side effects reported were dizziness, vertigo, incoordination, balance disorders, ataxia, tremor, somnolence, confusional state, attention disorders, thought disorders, euphoria, asthenia and fatigue **(Zaccara *et al.,* 2011)** .

Impaired vision

There have also been reports of visual adverse events including loss of vision, blurred vision or other changes in visual acuity, most of which were transient **(Zaccara *et al.*, 2011).**

Congestive heart failure

Post-marketing reports of congestive heart failure have been reported in some patients treated with pregabalin (Lyrica). These effects are mainly observed during pregabalin treatment for neuropathic pain in elderly patients with impaired cardiovascular function. Pregabalin should be used with caution in these patients. This adverse effect may disappear when pregabalin is discontinued **(Hickey and Thomas, 2011).**

Respiratory depression

Cases of severe respiratory depression have been reported in association with the use of pregabalin. Patients with impaired respiratory function, respiratory or neurological disease, renal insufficiency, or who are taking central nervous system depressant drugs in combination, particularly opioids, as well as the elderly, may be at greater risk of developing this serious side effect. Side effects were more frequent with increasing doses **(Zaccara *et al.*, 2011).**

Chapter 03:

Treatments

1. The aim of drug addiction treatment

Drug addiction treatment aims to help addicts stop compulsively seeking and using drugs. Treatment can take place in a variety of settings, take many different forms and last for different lengths of time.

As drug addiction is generally a chronic disorder characterised by occasional relapses, one-off short-term treatment is generally not enough. For many, treatment is a long-term process involving multiple interventions and regular monitoring **(Nora D *et al., 2011*).**

Drug addiction interventions aim to reduce recidivism by reducing drug use **(Goyette *et al.*, 2013).**

2. Principles of addiction care

In addition to the cognitive-behavioural approach, the most effective interventions include therapeutic communities **(Brochu and Plourde, 2012)**, medication-free maintenance treatments and motivational interviewing **(Holloway *et al.*, 2006).**

2.1. Cognitive-behavioural therapy

Cognitive-behavioural therapy (CBT) was developed as a method of preventing relapse. Cognitive-behavioural strategies are based on the theory that, in the development of maladaptive behavioural patterns such as drug addiction, learning processes play an essential role. People in CBT learn to identify and correct problem behaviours by applying a range of different skills that can be used to stop drug abuse and to solve a range of other problems that often accompany it. **(Caroll *et al.*, 2005).**

Research indicates that the skills acquired by individuals through cognitive-behavioural approaches persist after the end of treatment. **(Carroll *et al.*, 2008).**

2.2. Therapeutic communities

Therapeutic communities are also proving effective with drug addicts.

This approach is used in particular with people with severe substance use problems **(Casares-López *et al.*, 2013).**

Patients attend one or two counselling sessions each week, where they focus on improving family relationships, learning various skills to minimise drug use, receiving professional advice and developing new recreational activities and social networks **(Perry *et al.*, 2009).**

2.3 Medicines

Medication can help control withdrawal symptoms. However, medication is only the first step in treatment.

Medication may also be administered to restore normal brain function or curb the physiological need to use drugs. Medication varies according to the drugs involved and is administered under supervision as part of a treatment programme.

Like methadone, used alone to support opiate addicts.

It is thought that they are effective in reducing drug use problems **(Dolan *et al.*, 2005).**

2.4. Motivational interviewing

When applied to drug addiction, the motivational approach aims to help the addict recognise his or her current or potential problems caused by drug use and to encourage him or her to do something about them. The main objective is to increase the patient's intrinsic motivation to change. Motivation to change is the set of attitudes and behavioural dispositions conducive to changing drug-taking behaviour **(Miller *et al.*, 2006).**

This approach aims to evoke rapid, internally motivated change, rather than guiding the patient step by step through the recovery process. This therapy consists of an initial assessment, followed by two to four individual treatment sessions with a therapist.

In subsequent sessions, the therapist monitors change, reviews the withdrawal strategies used and continues to encourage commitment to change or sustained abstinence.

Patients are sometimes encouraged to bring a significant other to the sessions **(Miller *et al.*, 2003).**

In general, ME appears to be more effective in engaging drug users in treatment than in producing changes in drug use **(Baker *et al.*, 2002).**

3. The effectiveness of drug addiction treatment

As well as stopping drug abuse, the aim of treatment is to restore people to productive functioning in the family, workplace and community. According to research that follows people in treatment over long periods, most people who enter and remain in treatment stop using drugs, reduce their criminal activity and improve their occupational, social and psychological functioning. However, individual treatment outcomes depend on the extent and nature of the patient's problems, the appropriateness of the treatment and related services used to address those problems, and the quality of the interaction between the patient and his or her care providers **(Nora D *et al.*, 2011).**

Practical part

1. Issues

Drug addiction is a state of periodic or chronic intoxication caused by the repeated consumption of hallucinogens (psychotropic substances or medicines) at high doses, which inevitably entails health risks. This leads us to ask the following questions:

- What are the physical and social characteristics of drug users in the Tébessa region?
- What are the causes of this "drug addiction" phenomenon in the region?
- Which substances are most commonly used and what are their effects?
- What are the most common treatments for drug addiction?

2. Objective

The aim of this work is to carry out an epidemiological study of drug addiction in the Wilaya of Tébessa.

- Find out why hallucinogen use is on the rise and who the target population for drug addiction is.
- To show the effects of these substances on the health of drug users.

3. Study area

a. Location

According to **Abid (2015)**, the Wilaya of Tébessa, located in the far east of the country, covers an area of 13,878 km2 and had an estimated population of 800,000 in 2014, an average density of 58 inhabitants per km2. The capital of the wilaya, Tébessa, an ancient Roman city called Theveste, is located 40 km from the Tunisian border .

The wilaya of Tébessa is bordered: to the north, by the wilaya of Souk Ahras; to the east, by Tunisia; to the west, by the wilayas of Khenchela and Oum El Bouaghi; to the south, by the wilaya of El Oued **(Abid, 2015).**

It is made up of several geographical zones: to the north, the Tébessa mountains, which form part of the Atlas Mountains, the High Plateaux and the High Plains; to the south, the Saharan domain made up of a Saharan plateau **(Abid, 2015).**

The road network comprises 565km of national roads, 418km of wilaya roads and 1200km of communal roads. In economic terms, Tébessa has a threefold vocation: mining: exploitation of the Ouenza and Boukhadra iron mines and phosphate deposits, making it the country's leading iron and phosphate producer; agro-pastoral and industrial: establishment of cement works, flour mills, etc **(Abid, 2015).**

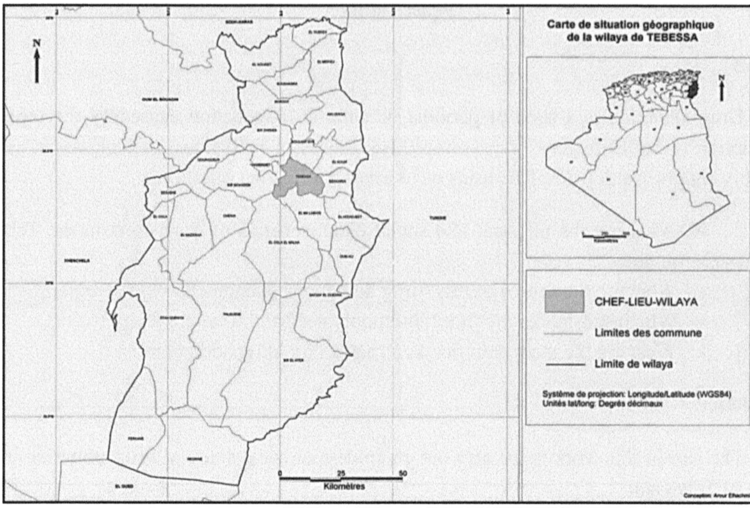

Figure 19: Map of Tébessa (communes and Daïras) **[Website 02]**.

The wilaya of Tébessa has 12 daïras: 1. Tébessa, 2. El Kouif, 3. Morsott, 4. El Ma Labiodh, 5. El Aouinet, 6. Ouenza, 7. Bir Mokkadem, 8. Bir El-Ater, 9. El Ogla, 10. Oum Ali, 11. Negrine and 12. Cheria **(ANIREF, 2015).**

b. Situation demographics

The wilaya's population is estimated at 693,671, giving an average density of 46 inhabitants per Km² **(ANIRF, 2015)**.

- Working population 166,457.
- Population 133,400.

4. Study methodology

4.1. Data collection

The sources of the data used for the study were randomly selected, at the Intermediate Centre for the Health of Drug Addicts in the Wilaya of Tébessa and outside the centre. It is reformed by a questionnaire for drug addicts in the form of papers and developed over one week (05 March 2023 to 09 March 2023).

4.2 Type and period of study

Our research study is descriptive, consulting and collecting data over a period of one and a half months (07 March 2023 to 16 April 2023).

4.3. Location

This work was carried out at the Centre Intermédiaire de Santé des toxicomanes (CIST) in Tébessa. The centre, which has been operational since 2010, has 5 offices, 2 waiting rooms, an archive, a department head's office, a group room, a treatment room and a laboratory. The staff working directly with drug addicts include a general practitioner specialising in addictology, two psychologists, four nurses and an admirer.

Figure 20 : Intermediate health centre for drug addicts -Tébessa-.

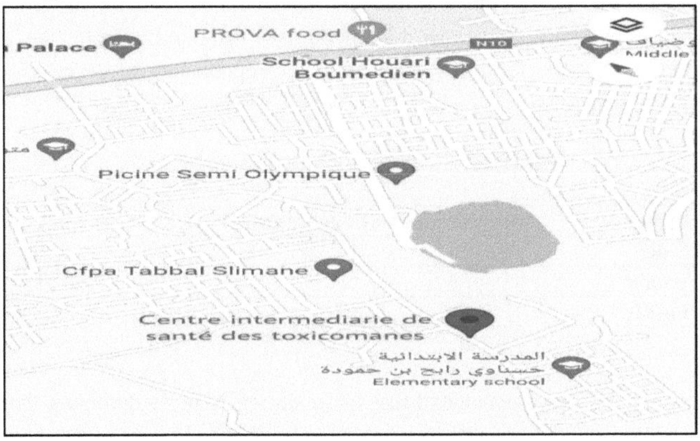

Figure 21: Location of the Tébessa CIST **(Google maps, 2023).**

4.4. Study population

The population studied was drug addicts. Our sample consisted of 60 patients (49 men and 11 women), all of whom completed the questionnaire **(Appendix 1).** Their ages ranged from under 15 to over 35 years, and their educational levels ranged from primary to secondary and university.

4.5. Study variable

The epidemiological parameters analysed in this study are :

- Age.
- Sex.
- Professional situation.
- Family situation.
- Substances consumed.

4.6. Data processing and analysis

The data was processed on a computer using Microsoft Excel 2016 software for the tables and statistical analyses. The results were compiled and presented in the form of histograms.

5. Results and discussions

Based on the information gathered, a descriptive and analytical statistical analysis was carried out. The results were as follows:

5.1. Personal characteristics of patients

5.1.1. Breakdown of drug users by sex

Graph (22) and table (03) show the repetition of drug addiction by sex. According to the results, men are more affected by drug addiction, represented by 49 cases, or 82%, and women by 11 cases, or 18%.

Table 03: Breakdown of drug users by sex.

Gender	Number of drug addicts	Percentage
Male	49	82
Female	11	18
Total	60	100

To this end, we have concluded that the male sex is more dominant than the female sex and we obtained this result because men responded more than women to our questionnaire.

Almost all the patients were male, i.e. 57 men (71%) and 23 women (29%). These data corroborate the work of **Nedjari (2009),** who reported higher prevalence rates in Algeria, of the order of 94% males and 6% females. Similarly, **Azzeddine (2012)** found a prevalence rate of 97.5% males and 2.5% females in Algeria. In another study, **Ghomari and Oussahla (2019)** reported a prevalence rate among professional drivers of 99.5% for men and 0.5% for women.

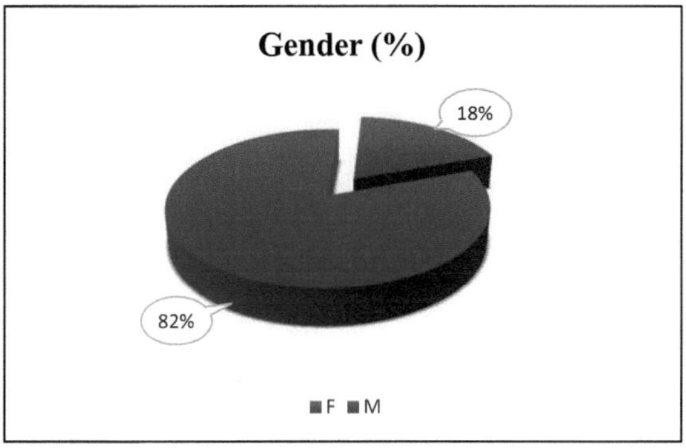

Figure 22: Breakdown of drug users by sex.

5.1.2. Breakdown of drug users by age

Graph (23) and table (04) show the distribution of drug addicts by age. Our results show that the majority of cases affect people in the (20-25) age group, with 29 people being the most vulnerable, since statistically they constitute the dominant and most active sections of the population, whereas the (-15) age group is the least affected by drug addiction, with only one case.

Table 04: Breakdown of drug users by age .

Age	Number of drug addicts	Percentage
(-15)	1	2
(15-20)	14	23
(20-25)	29	48
(25-30)	10	17
(+35)	6	10
Total	60	100

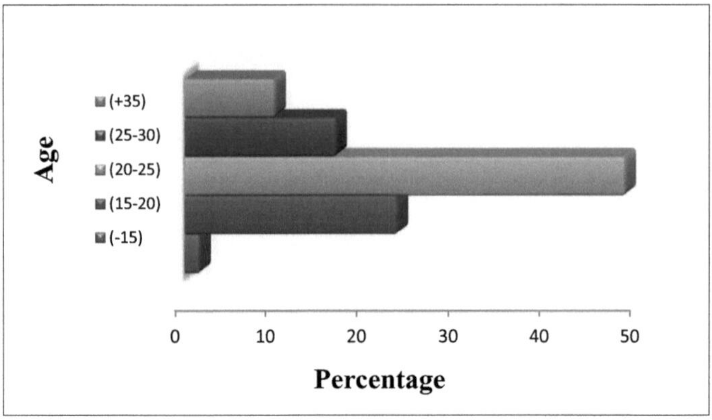

Figure 23: Breakdown of drug users by age.

The population most concerned is in the 20-25 age group (48%), which means that young people of a certain age are looking for a different experience, a new sensation and a way of coping with a difficult situation.

Therefore, the results obtained are the same as those found by **Kaddour (2015)** in Tlemcen, who noted that the patients who consume the most are aged (18-24) years (45%) and (25-34) years (32%). And also for the results of **Touami *et al*, (2018)** in Mostaganem, the most targeted age is (18-33) years (45%).

5.1.3 Distribution of patients by educational level

Table (05) and **Figure 24** show that the educational level of patients who take these hallucinogens is (40%) university level and the same as that of high school patients (40%), while (17%) of patients have an average level, and a very small percentage have a primary level (3%).

Table 05: Breakdown of patients by educational level.

School level	Number of patients	Percentage
Primary	2	3
Medium	10	17
Lycée	24	40
University	24	40
Total	60	100

The use of drugs is more is more widespread among subjects who have both university and secondary levels of (40%) : Hallucinogens are abundant in institutions, study depression and too many gatherings.

On the other hand, **Kaddour (2015)** reported that 46% of patients had secondary education. The same finding was made by **Azzeddine (2012)** in Algeria, with 48.75% having an intermediate level and 17.5% a secondary level.

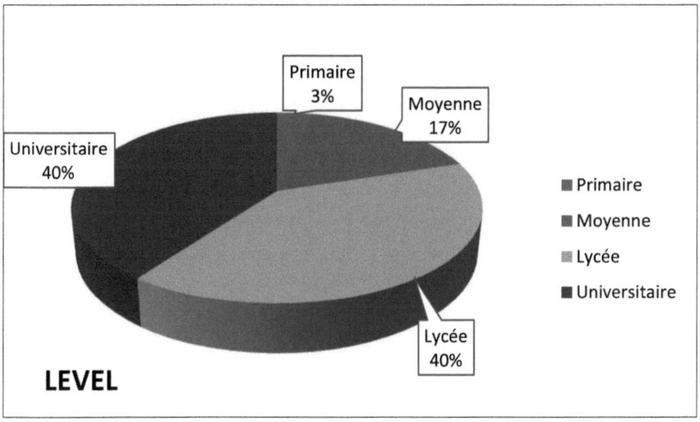

Figure 24: Distribution of patients by educational level

5.1.4. Breakdown of patients by social status

Table (06) and Figure (25) show that the highest percentage of drug users (addicts) are single (85%), followed by married (13%), while divorced patients account for only 2%.

Table 06: Breakdown of patients by social status

The social state	Number of patients	Frequency
Single	51	85
Married	8	13
Divorced	1	2
Total	60	100

Bad friends and emptiness are the main causes of drug addiction.

They are similar to those reported by **Ammari *et al* (2021)**, where most consumers were 84% single, compared with 11% married and 4% divorced.

Even **Mebarki (2015)** is also He found that the majority were single 56%.

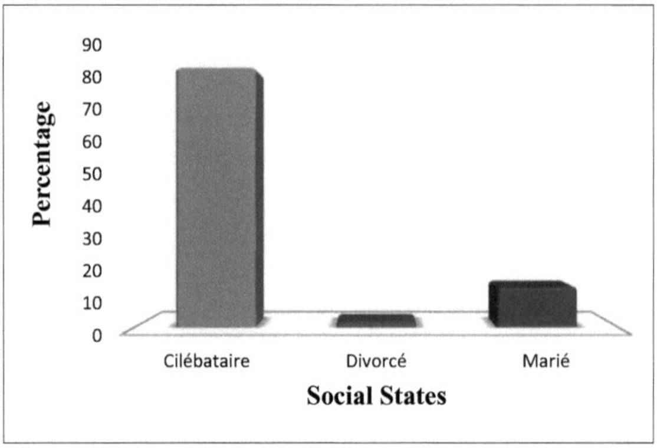

Figure 25: Distribution of patients according to social status.

5.1.5. Breakdown of patients by type of length of stay

Table (07) and Figure (26) show that the highest percentage of these substances is consumed in the family home, estimated at (80%), followed by private housing with a percentage of (13%), and then a small percentage in rented accommodation with (7%).

Table 07: Breakdown of a drug addict according to housing situation.

Long	Number of patients	Frequency
Family	48	80
Private	4	7
Rental	8	13
Total	60	100

These results show that most drug users are still adolescents living with their families. Our results are comparable to those of **Ammari *et al* (2021)** the highest percentage of these substances is consumed in the family home (82%), followed by private housing a percentage of (15%), then a small percentage in rentals with (3%).

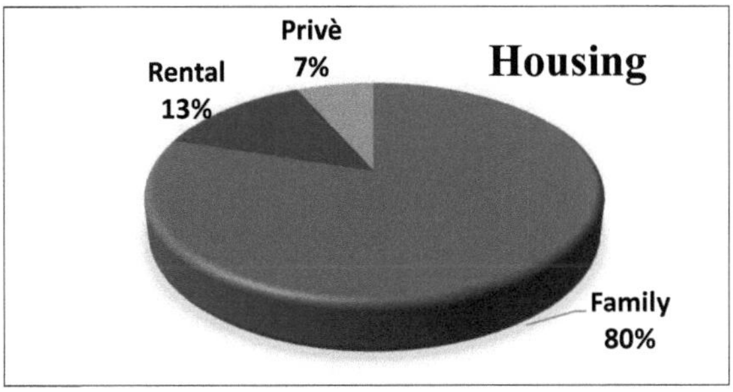

Figure 26: Breakdown of patients by type of accommodation.

5.2. Information on the use of hallucinogens

5.2.1. Breakdown of patients according to time spent taking hallucinogens

It should be noted that drug users are estimated to take drugs throughout the day (38%), followed by morning (35%), evening (19%) and night (8%) drug users.

Table 08: Temporal distribution of hallucinogen use.

Time to take drugs	Number of patients	Frequency (%)
All the time	18	38
The morning	17	35
In the evening	9	19
The night	4	8
Total	60	100

Drugs can be consumed at any time of the day, with the availability of each substance and the habits and preferences of each addict determining when it can be used during the day.

These results are contrary to those reported by **Ammari et al (2021),** where most drug users consumed in the evening (44%), followed by the afternoon (31%), then the morning (25%).

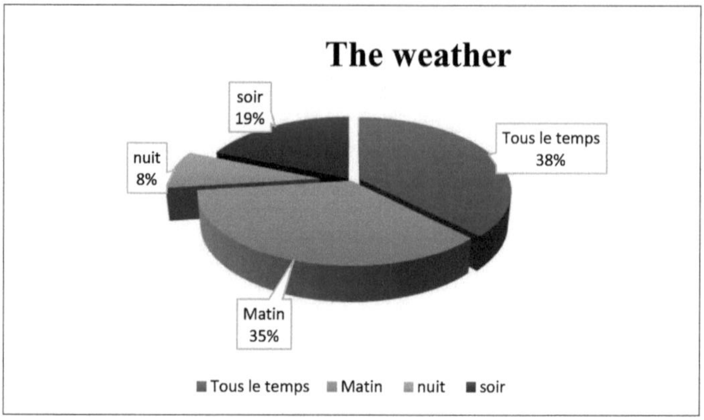

Figure 27: Breakdown of drug users by time spent taking hallucinogens.

5.2.2. Breakdown of patients by type of hallucinogen

Table (09) and Figure (28) show that lyrica (Pregabalin) with a percentage of (59%) is the most widely used psychoactive substance in terms of addiction, followed by tramadol (18%), followed by users of other substances at a rate of (16.4%) and on the other hand patients who use kietyl represent only (6.6%).

Table 09: Breakdown of patients according to type of hallucinogen.

Type of substance	Number of patients	Frequency
Lyrica	35	59
Tramadol	11	18
Kietyl	4	6.6
Other substance	10	16.4
Total	60	100

These results show that Lyrica (pregabalin) is not too expensive and has become widely available.

They are similar to those reported by **Ammari *et al* (2021)**, where most users consume pregabalin as a percentage (68%), other drugs (18%), followed by kietyl a percentage (8%) and Tramadol (6%).

Similarly, **Ghomari** and **Oussahla (2019)** reported that pregabalin (Lyrica) was the most commonly used drug after cannabis a percentage (3.7%), followed by kietyl (2.8%), cocaine (2.3%), and Amphetamines (Ecstasy) (1.4%).

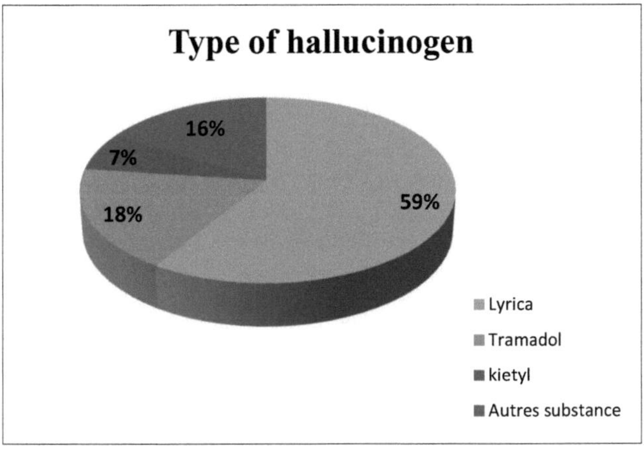

Figure 28: Proportion of hallucinogens most frequently used.

5.2.3. Breakdown of drug users by mode of consumption

Graph (29) and table (10) show the recurrence of drug users by mode of consumption. According to the results obtained, we noted :

- Hallucinogenic substances that are consumed orally, with 70% of the total consumption.
- By intravenous route in very small percentage 10%.
- There are also several ways of consuming hallucinogens, but only a small percentage (20%).

Table 10: Breakdown of drug users by mode of consumption.

Consumption mode	Number of drug addicts	Percentage
Tablet	42	70
Injection	6	10
Other	12	20
Total	60	100

The ease with which tablets can be taken, their cheaper price and the fact that they are more widely available in our society have all contributed to these results.

Comparison of our results with other studies on the same subject in Oued-Sauf **Ammari *et al,* (2021)** show that the majority of patients (98%) take the drug in pill form, but very few take it in the form of an injection or suppository (01%).

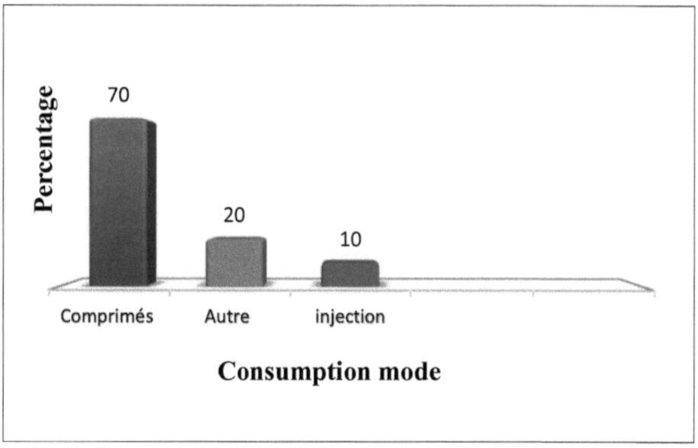

Figure 29: Breakdown of drug users by mode of consumption.

5.2.4. Breakdown of patients by drug price

The graph **(Figure 30)** and table (11) show that most drug users say that hallucinogens are readily available (68%), while 20% of patients say that these substances are expensive, and the remaining 12% say that drug prices are too high.

Table 11: The price of hallucinogens

The price of hallucinogens	Number of patients	Frequency (%)
Affordable	41	68
Dear	12	20
Too expensive	7	12
Total	60	100

This result was confirmed by those mentioned in **(Ammari et *al.*, 2021),** where the majority of drug users (49%) indicated that the drug was cheaper, followed by (40%) indicated that it was more expensive, On the other hand (11%), they stated that the price was too high.

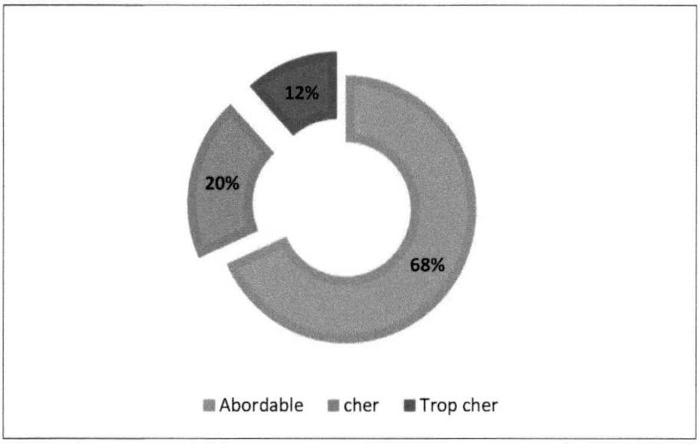

Figure 30: Distribution of patients by drug price.

5.2.5 Breakdown of drug users by method of obtaining hallucinogens

According to the documented results, the majority of drug users obtained their drugs from vendors, which we noted at a high rate (85%), followed by non-prescription estimates (10%) , and from friends (3%), and on the other hand (2%) drug users use other methods to obtain hallucinogens.

Table 12: Breakdown of drug users by method of obtaining hallucinogens.

Method of production	Number of drug addicts	Frequency (%).
Sales	51	85
Over-the-counter purchases (psychoactive drugs)	6	10
Our friends	2	3
Other method	1	2
Total	60	100

The results are similar to those of the study by **Ammari *et al*, (2021)** in which it is noted that the most popular method used by drug addicts to buy the drug is through sellers estimated at (60%), followed by those who buy SPAs without a prescription being used by (14%) of drug addicts, and the remainder use other methods to obtain it.

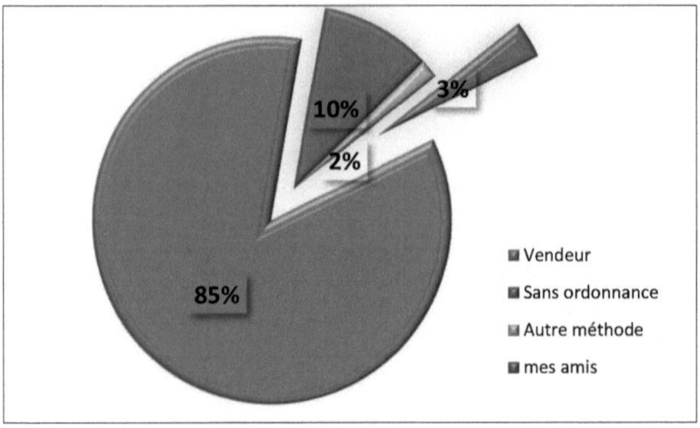

Figure 31: Breakdown of drug users by method of purchase of hallucinogens.

5.2.6. Breakdown of patients by reason for taking hallucinogens

In this study, we find that the main cause of addiction among patients is attributed to a bad environment, accounting for an estimated 33%. This is followed by addicts with family problems, who account for 23% of cases. On the other hand, 18% of individuals become addicted because of problems related to studies and unemployment, while 17% are addicted for the same reasons. Finally, 8 of people take drugs because of unfavourable social conditions.

Table 13: Breakdown of patients according to the reasons for taking hallucinogens.

Reasons for taking hallucinogens	Number of drug addicts	Frequency (%)
Bad friends	20	33
Family problems	14	23
Study problems	11	18
Unemployment	10	17
Social conditions	5	8
Total	60	100

It is not easy to recognise the reasons why certain drugs are taken by drug addicts themselves, who find it difficult to describe the causes of their consumption.

The most frequently reported reasons were: bad friends 51%, social conditions 20%, followed by family problems 16%, work pressure and unemployment estimated at 12%, and learning problems **Ammari et *al.*, (2021).**

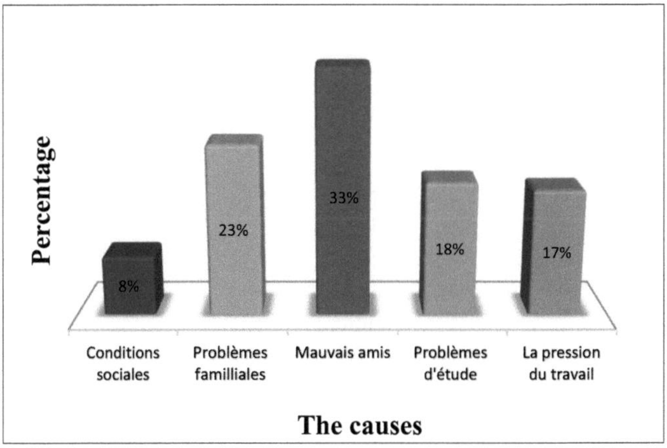

Figure 32: Breakdown of patients by reason for taking hallucinogens.

5.2.7. Breakdown of patients by number of doses taken daily

Table (14) and figure (33) show that a significant rate of 2 doses / day percentage (35%), followed by >> 5 doses / day (25%) and a single dose with a rate of (15%), and other doses including (13%) and (7%), followed (5%).

Table 14: Breakdown of patients by consumption dose.

Number of doses taken daily	Number of patients	Frequency (%)
1	9	15
2	21	35
3	8	13
4	3	5
5	4	7
>>5	15	25
Total	60	100

These results show that some drug addicts are not satisfied with one dose because it doesn't give them the euphoria they need, so they consume higher and higher doses of a drug to obtain the same effects.

They are similar to those reported by **Ammari *et al* (2021)**, with a high rate of 2 doses/day (27%), followed by 3 doses/day (25%) and more than 5 doses (20%), and other doses including 15% and 13%, respectively.

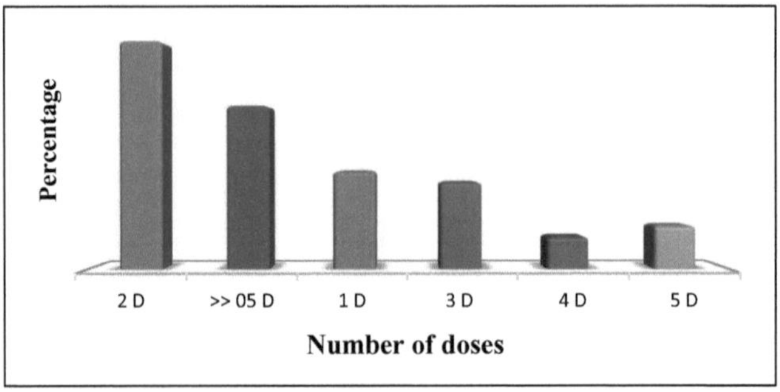

Figure 33: Distribution of patients according to the number of doses taken daily.

5.2.8. Breakdown of drug use by year

In the graph in the following figure, we see a high percentage (50%) of drug use in (2016-2017) and lower and similar percentages in (2018-2020) of 23%, (2021-2022) of 14% and in (2010-2015) of 9%. In contrast, the use of these drugs fell by 4% in (2000-2007).

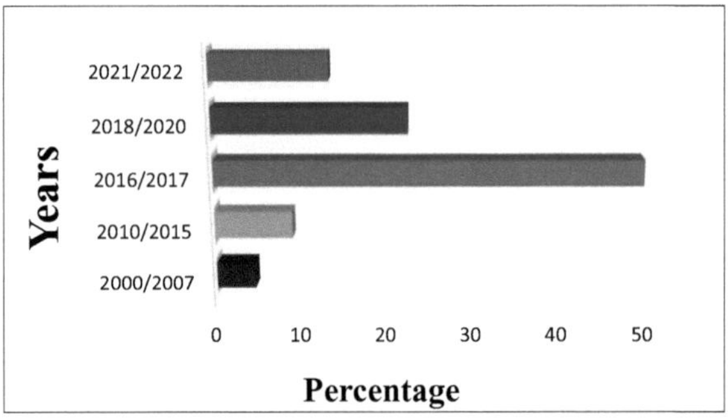

Figure 34: Breakdown of drug use by year.

These results indicate that during this period (2016-2017) the majority of drug addicts were prey to the consumption of hallucinogens, which became more widespread due to the availability of these substances and the fragility of the economic and social infrastructure.

According to **M. Benhalla (2018)**, Algeria is a drug-consuming country and psychotropic substances are consumed much more during the years (2016-2018).

5.2.9. Breakdown of patients by age of onset of drug use

Table (15) and figure (35) show that most drug users report their first contact with hallucinogens at the age of 20 to 25 with a percentage (62%), followed by Under 15 (25%) and 20 to 25 (8%) and 5% from 25 to 30.

Table 15: Breakdown of patients by age of onset of drug use

Initial consumption	Number of patients	Frequency (%)
Under 15s	15	25
15 to 20 years	37	62
20 to 25 years	5	8
25 to 30 years	3	5
Total	60	100

These results translate into : Adolescence is one of the main causes of drug addiction, even the absence of family control, also bad friends.

Our results are comparable to those of **Mebark (2015)** indicates that The average age of onset was 18 ± 06 years. The peak age of onset was between 16 and 26 years.

Similarly, **Ammari *et al* (2021)** The majority of patients reported their first use of psychotropic substances; in the adolescent phase of 15 to 20 years (47%), followed by 21 to 25 years (27%) and less than 15 years (14%). The lowest prevalence was between the ages of 25 and 30 (09%) and 30 and over (03%).

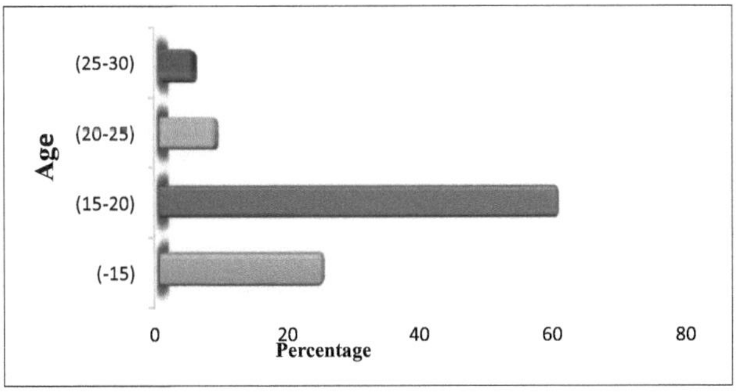

Figure 35: Breakdown of patients by age of onset of drug use.

5.2.10. Distribution of patients according to the effect of hallucinogenic substances on the body

According to the breakdown of effects on the body, 33% of consumers suffer from neurological problems, while 28% suffer from respiratory problems. A further 22% suffer from digestive problems, and an estimated 17% have heart problems.

Table 16: Undesirable effects on the human body.

The harmful effects of hallucinogens	Number of patients	Frequency (%)
Nervous system disorder	20	33
Respiratory disorder	17	28
Digestive problems	13	22
Heart problems	10	17
Total	60	100

We note that the results are similar to the study by **Bourouis et *al*, (2019),** where the most frequent clinical signs observed in patients are: nervousness (72.72%), drowsiness (36.36%), stress and paleness (45.45%), headaches, weakness, slowness, dizziness (90.90%), weight loss (81.81%), insomnia (80%) and dehydration (36.36%).

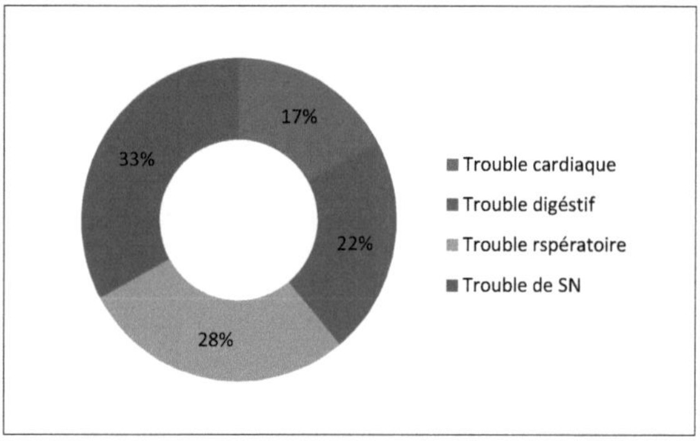

Figure 36: Distribution of drug users according to harmful effects on the body.

5.2.11. Breakdown of smoking drug addicts

The table and graph show the patients' smoking habits. Of the registered cases, the majority of 49 (82%) are smokers, while only 11 (18%) of the registered cases do not smoke.

Table 17: Breakdown of drug addicts by smoking association.

Smokers	Number of drug addicts	Percentage
Yes	49	82
No	11	18
Total	60	100

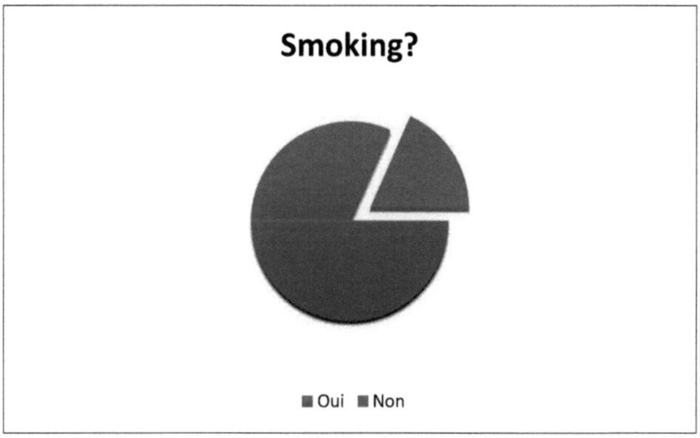

Figure 37 : Breakdown of drug users by smoking association.

The majority of addicts prefer the combination of tobacco and cigarettes, because cigarettes make them feel relaxed and give them a sense of energy. Smokers often see it as a friend that helps them get through difficult times. Boredom and loneliness can also lead some people to smoke.

According to the National Institute of Public Health in Tunisia Tobacco consumption in Tunisia was reported by 30.4% of the subjects surveyed. It concerned 55.6% of men, but only 5.2% of women.

In 2017, according to the Santé publique France Barometer, daily smoking among 18-75 year-olds (standardised rates) varied from 21.3% to 32.2%.

5.2.12. Breakdown of drug users by alcoholism association

Alcohol consumption is shown in table (18) and graph (38). The results obtained show that people who say no represent the majority with 38 (63%) and people who say yes are represented by alcoholics 37% with 22 cases.

Table 18: Breakdown of drug addicts by alcoholism association.

Alcoholic	Number of drug addicts	Percentage
Yes	22	37
No	38	63
Total	60	100

The consumer uses this type of combination to amplify the pleasurable effects, but may also use other combinations to reduce the unpleasant effects experienced, such as anxiety or drowsiness, or simply to cut costs. Despite this, it is more common to use other substances than alcohol.

In contrast, nearly 8.83%, or 6 patients in our population, associated alcohol consumption with drug use. **Toe (2011)** found that alcohol consumption was most often associated with cannabis. This association was also shown by **Traore (2004).**

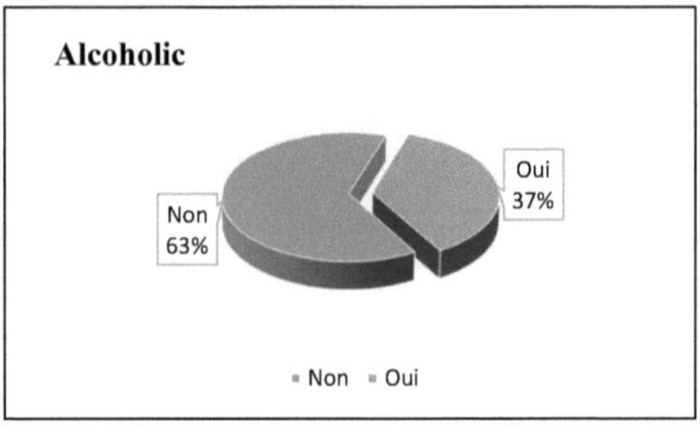

Figure 38 : Breakdown of drug addicts by alcoholism association.

5.3. Diagnostic features

5.3.1. Carrying out various daily activities after using hallucinogens

Table (19) and figure (39) show that most drug addicts (95%) can carry out their daily activities normally. The remaining (5%) cannot carry out their activities.

Table 19: Breakdown of activities after using hallucinogens

Activity after using hallucinogens	Number of patients	Frequency (%)
Yes	57	95
No	3	5
Total	60	100

These results show that most of the respondents take hallucinogens to feel active and vital. Most of them can't do anything if they don't.

Ammari et al (2021) found that most respondents (75%) were able to carry out their daily activities. The remaining 25% were unable to carry out their activities.

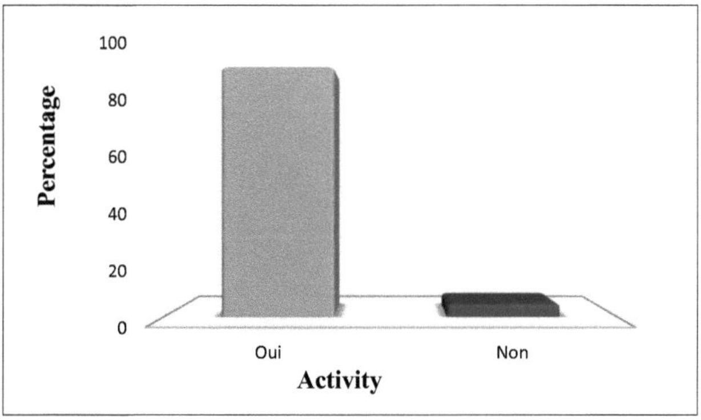

Figure 39: Breakdown of patients' activities after using hallucinogens.

5.3.2 Distribution of patients according to feelings after taking hallucinogens

Table (20) and figure (40) show that most consumers feel comfort/happiness at its rate is up (50%) and activity/vitality at percentages (36.7%), in contrast patients feel tired/exhausted (10%), followed by a very low rate of patients feel sadness and depression (3.3%).

Table 20: Distribution of patients according to feelings after taking hallucinogens.

Feeling after taking the medicine	Number of patients	Frequency (%)
Comfort / happiness	30	50
Activity / vitality	22	36.7
Tired / exhausted	6	10
Sad / depression	2	3.3
Total	60	100

Most addicts use drugs for happiness, comfort and activity. Some of them cannot enjoy themselves or even do their work or anything else without taking them.

They are similar to those mentioned by **Ammari *et al* (2021),** where the majority of consumers feel happy (54%), active and vital (36%) after taking medication and in a few percentages they feel sad and depressed (03%), tired and exhausted (07%).

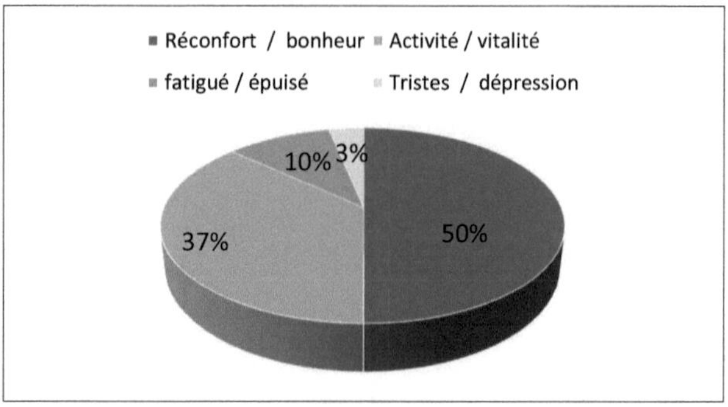

Figure 40: Distribution of patients according to feelings after taking hallucinogens.

5.3.3. Breakdown of patients by health problems (psychological, organic)

Table (21) and figure (41) show that the highest percentage of addicts have psychological problems (45%), followed by addicts with no problems (42%) . However, patients suffering from organic problems represent only (13%).

Table 21: Breakdown of drug users by type of problem.

Type of problem	Number of patients	Frequency (%)
Psychic	27	45
No	25	42
Organic	8	13
Total	60	100

This result translates into the frequency of psychiatric comorbidity in drug-using subjects is considered by all authors to be particularly high. People with psychiatric disorders may be more likely to develop a psychoactive substance, as they may use the substance for their psychiatric symptoms.

They are similar to those mentioned by **(Ammari *et al.*, 2021)**, where we note that the largest percentage of drug addicts suffer from a psychological problem, estimated at (55%), followed by those who suffer from no problem at all with (41%) and then a very low percentage for those who suffer from an organic problem, estimated at (3%).

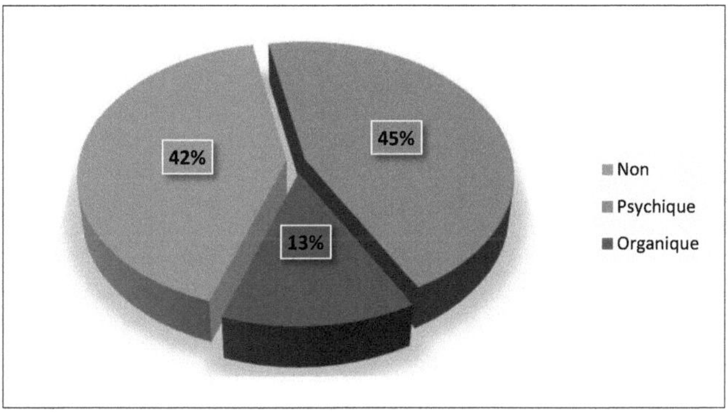

Figure 41: Breakdown of patients by type of problem.

5.3.4. Breakdown of drug addicts by state of withdrawal from hallucinogens

Table (22) and figure (42) show that the percentage of patients who experience anxiety, stress and loss of control when they run out of hallucinogens is the highest (43%). Conversely, (38%) of drug addicts experience anxiety and stress, while (12%) of patients experience a loss of control. On the other hand, only (7%) of patients do not suffer from a deficiency of these substances.

Table 22: Breakdown of drug users according to craving.

Drug addiction in the event of withdrawal	Number of drug addicts	Frequency (%)
Anxiety and stress/loss of control	26	43
Anxiety and stress	24	38
Loss of control	7	12
Normal state	3	7
Total	60	100

When drug use is reduced, the body reacts with signs of withdrawal, sometimes nervous disorders (anxiety, depression) or hallucinations.

Taking the drug reduces or even eliminates these withdrawal symptoms, encouraging the addict to continue using it regularly.

We obtained these results in an intermediate drug treatment centre, where we observed patients' behaviour when they were not using drugs.

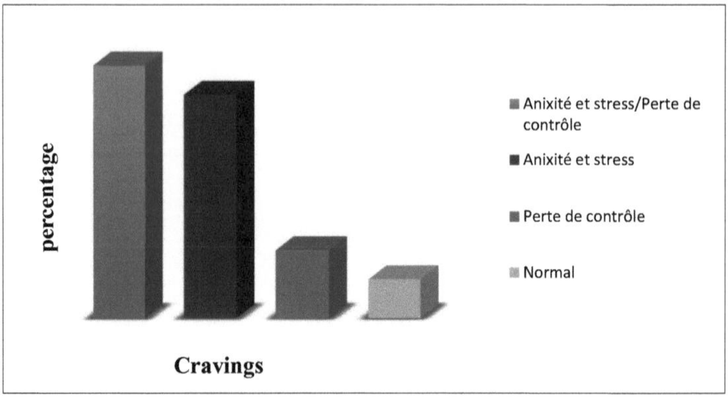

Figure 42: Breakdown of drug addicts by state of withdrawal from hallucinogens.

5.3.5. Distribution of patients according to their views on drug use

(67%) of addicts have a negative view of addiction, (25%) have a normal view and (8%) have a favourable view of addiction, according to the breakdown of addicts' views on addiction.

Table 23: Breakdown of patients according to their views.

The opinions of drug addicts	Number of patients	Frequency (%)
Negative view	40	67
Normal vision	15	25
A positive outlook	5	8
Total	60	100

It is important to note that the experiences and opinions of addicts can vary considerably from one person to another. Some people may see their addiction as a burden, while others may find comfort or escape in their use of substances.

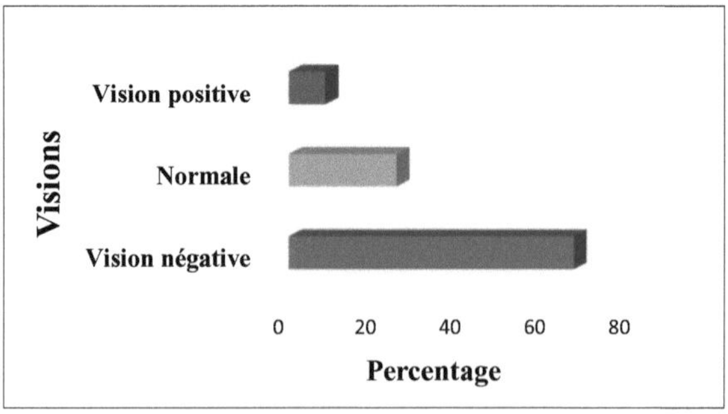

Figure 43: Distribution of patients according to their views on drug use.

5.3.6. Thinking about stopping consumption

Table (24) and Figure 44 show that the highest percentage of people thinking about stopping using drugs is estimated at (88%), followed by the opposite response with a percentage of (12%).

Table 24: Representation of patients thinking about quitting.

Thinking about stopping drugs	Number of patients	Frequency (%)
Yes	53	88
No	7	12
Total	60	100

These results are the cause of: the respondents are sure that these hallucinogens have negatively affected their health and their relationships.

They are similar to those reported by **Ammari *et al* (2021),** finding that the study population thought to stop were yes estimated at (87%), followed by the opposite response with a percentage of (13%).

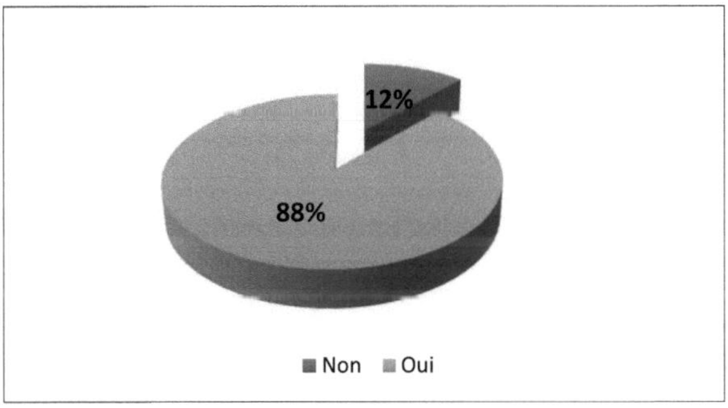

Figure 44: Distribution of patients who are thinking about stopping consumption.

5.3.7. How did the addict think of uprooting these hallucinogens?

Table (25) and **figure (45)** show that the highest percentage of people who decide to stop taking these drugs on their own is estimated at (50%), followed by (27%) going to the addiction centre and (18%) to psychologists.

Table 25: Representation of patients according to their thoughts about uprooting these hallucinogens.

The thought of stopping these Hallucinogens	**Number of patients**	**Frequency (%)**
Stop alone	33	50
Drug addiction centre	16	27
Psychologist	11	18
Total	60	100

According to our results, it has been established that certain hallucinogenic substances are not addictive, which means that addicts can stop using them without medical assistance.

However, other substances are addictive and require medical treatment under the supervision of psychologists and psychiatrists.

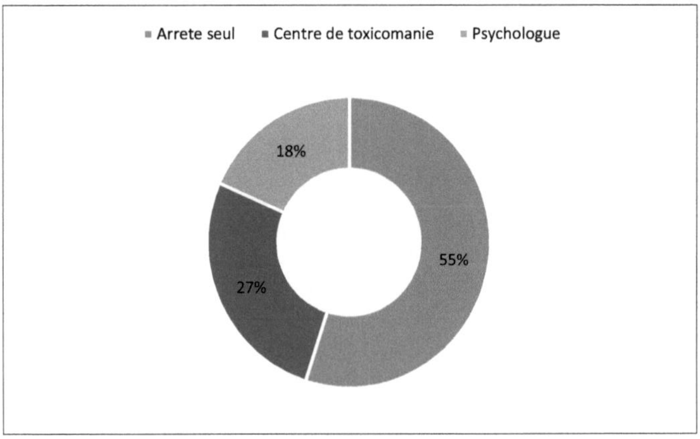

Figure 45: Distribution of patients according to their thoughts about uprooting these hallucinogens.

Conclusion

In conclusion, this dissertation has taken an in-depth look at the phenomenon of drug addiction, highlighting the scale of the global problem of substance dependence and emphasising the importance of a holistic approach to tackling this crisis. In Algeria, drug addiction has become a serious problem due to the growing increase, in terms of both quantity and quality, of this scourge, which mainly affects an increasingly young population. The regular use of a variety of drugs, including cannabis and psychotropic medicines, is a daily reality, with the emergence of a clientele of young people addicted to hard drugs.

Our detailed and rigorous analysis of the risk factors associated with drug use revealed that the majority of the patients we met were multiple users of psychoactive substances, such as tobacco, alcohol, amphetamines, cannabis, ecstasy and Lyrica®. They reported adverse health effects, socio-economic problems and dependence. Most of these multiple users were single, well educated and well paid , and had started using at an early age, with underlying psychological problems.

The results of this study carried out at the Tébessa intermediate health centre for drug addicts (CIST) can be summarised as follows:

- The profile of the patients is characterised by their young age as young adults [21-25 years].
- The majority of participants are single men, in higher proportions than women.
- Most patients are well educated and receive satisfactory remuneration.
- Pregabalin was identified as the most commonly used substance, followed by Tramadol.
- The consumption of hallucinogens has undesirable effects on the bodies of the individuals concerned.
- Most patients want to stop taking hallucinogens.

During our research, we encountered difficulties in communicating with drug users and obtaining the necessary information. In addition, we wanted to diagnose the drug users in greater depth by carrying out analyses, but we encountered material constraints, in particular the lack of reagents required to carry out drug screening tests such as the Drogue test or the Narco test.

In perspectives , this work has also highlighted certain limitations and gaps in the existing literature, paving the way for new research questions. Further in-depth studies are needed to better understand this scourge and to propose effective solutions.

Within specific limits and constraints, this dissertation makes a significant contribution to the field of addiction. It offers an in-depth analysis and new perspectives on addiction, representing an important contribution to academic research in this field. We sincerely hope that this dissertation will be a valuable resource for researchers, practitioners and policy makers interested in this phenomenon and that it will stimulate future research to advance our understanding and actions in the fight against addiction.

References

- **Alain.** M & Grégory. P. (2019). Hallucinogens and synthetic drugs in addictology. P219-232.
- **Alain M.** (2015). Hallucinogens and synthetic drugs in addictology. P170-181.
- **Amar, M. B & Léonard, L.** (2002). Psychotropic drugs: pharmacology and addiction,9. P5-8
- **Audrey, D. Rosse, L. Agathe** (2010). Toxicomanie : Rencontre d'un individu, d'un produit, d'une société, P5.
- **AZZEDDINE R.** (2012). Schizophrenia cannabis addiction comorbidity. University of Oran 1- Ahmed Ben Bella, Doctoral thesis in medical sciences. P120
- **Ammari M, Harzallah C, Nagoudi H, Saighi L,** Survey on the consumption of hallucinogens and their effects on drug addicts (case of the wilaya of El Oued) (2022)
- **Baker A, Lewin T, Reichler H, Clancy R, Carr V, Garrett R, Terry M.** (2002). Evaluation of a motivational interview for substance use within psychiatric in-patient services. Addiction, 97(10). P1-10
- **Bensakhria. A.** (2015). Addiction natural and synthetic drugs. P3
- **Benyamina, A.** (2017). Addictions: From pleasure to addiction. P12-15
- **Bian, F., Li, Z., Offord, J., Davis, M. D., McCormick, J., Taylor, C. P., & Walker, L. C.** (2006). Calcium channel alpha2-delta type 1 subunit is the major binding protein for pregabalin in neocortex, hippocampus, amygdala, and spinal cord: An ex vivo autoradiographic study in alpha2-delta type 1 genetically modified mice. Brain Research, 1075(1), 68-80. https://doi.org/10.1016/j.brainres.2005.12.084.
- **Bockbrader, H. N., Wesche, D., Miller, R., Chapel, S., Janiczek, N., & Burger, P.** (2010). A comparison of the pharmacokinetics and pharmacodynamics of pregabalin and gabapentin. Clinical Pharmacokinetics. P13-21
- **BOUROUIS N., LAIMECHE A., CHOUKI S., LOURGUIOUII** (2021). Drug screening as part of detoxification treatment at the Tizi-Ouzou addiction teaching, research and treatment centre. Doctor of Pharmacy thesis. Mouloud Mammeri University. Faculty of Medicine TIZI OUZOU: P190.
- **Brochu, S., & Plourde, C.** (2012). L'offre de services aux adultes toxicomanes sous le coup de mesures judiciaires. L'integration des services en toxicomanie, P107.
- **Burnat, P., & Le Brumant-Payen, C.** (1992). Cocaine intoxication. Lyon Pharm, Vol 43, 149-156.

- **Burnat, P., Huart, B., Ceppa, F., & Pailler, FM.** (1996). Ecstasy: psychostimulant, hallucinogenic and toxic substance. Presse médicale (Paris, France: 1983*)*, *25* (26), 1208-1212.

- **Carl L. Hart, Joanne Csete, Don Habibi.** (2014). Methamphetamine-danger-exagerated. 38 p.

- **Carroll, K. M., & Onken, L. S. (2005).** Behavioral therapies for drug abuse. American Journal of Psychiatry, 162(8), 1452-1460.

- **Carroll, K. M., Ball, S. A., Martino, S., Nich, C., Babuscio, T. A., Nuro, K. F., & Rounsaville, B. J.** (2008). Computer-assisted delivery of cognitive-behavioral therapy for addiction: a randomized trial of CBT4CBT. American Journal ofPsychiatry, 165(7).

- **Casares-López, M. J., González-Menéndez, A., Festinger, D. S., Fernández-García, P., Fernández-Hermida, J. R., Secades, R., & Matejkowski, J.** (2013). Predictors of retention in a drug-free unit/substance abuse treatment in prison. International journal of law and psychiatry, 36(3-4).

- **Cassier, Vincent** (2014). Retrospective study of acute tramadol intoxications (2007-2013) based on data from the Bordeaux poison and toxicovigilance centre: impact of the withdrawal of dextropropoxyphene. Doctoral thesis. University of Bordeaux, P14.

- **CÉLINE RAY, M**. (2017). Reward circuit: what is it?

- **Chambon, O.** (2009). La Médecine psychédélique -Le pouvoir thérapeutique des hallucinogènes. Les Arènes.

- **Charles, FLEURENTIN.** (2015). Therapeutics using psychedelic psychotropic drugs. Doctoral thesis. UNIVERSITE DE LORRAINE, 34.

- **Chaumont, O.** (2018). Addiction, dependence and drugs: a study of the implementation of low-risk consumption rooms in the world and in France since 2016. PhD thesis. University of Bordeaux. 144p.

- **Ciciero, T.J. and M.S. Ellis** (2015). "Abuse-deterrent formulations and the prescription opioid abuse epidemic in the United States: lessons learned from oxycontin," **JAMA Psychiatry,** vol. 72, no. 5.p 424-430.

- **Cirriez, P.** (2015). Journal de Pharmacie de Belgique: Common illicit drugs used by drug addicts, P14.

- **Constance, H.** (2002). Drogues, médicaments, dépendances, Ecstasy P35-49.

- **From a group.** (1 January 2006). Drugs and dependence. Information book. Editions Insep. France, Pp36.

- **Degenhardt, L., Charlson, F., Ferrari, A., et al**. (2018). The global burden of disease attributable to alcohol and drug use in 195 countries and territories, 1990-2016: a systematic analysis for the Global Burden of Disease Study 2016. The Lancet Psychiatry, 5(12), 987-1012.

- **Demaret, I., Lemaître, A., & Ansseau, M.** (2013). L'héroïne. Revue médicale de Liège, Vol 68, P 5-6.

- **Devaux, Lucie** (2022). Tramadol use: between misuse and dependence. Pharmaceutical sciences. Doctoral thesis. Université de Picardie,50.

- **Devaux, Lucie** (2022). Tramadol use: between misuse and dependence. Pharmaceutical sciences. Doctoral thesis. Université de Picardie, 33-39.

- **DEVER, L.** (2015). Drugs and Treatments,38-48.

- **Dolan, K. A., Shearer, J., White, B., Zhou, J., Kaldor, J., & Wodak, A. D.** (2005). Four-year follow-up of imprisoned male heroin users and methadone treatment: mortality, re-incarceration and hepatitis C infection. Addiction, 100(6).

- **Dooley, D. J. Donovan, C. M. Meder, W. P., & Whetzel, S. Z.** (2002). Preferential action of gabapentin and pregabalin at P/Q-type voltage-sensitive calcium channels: inhibition, Vol 45,171-190.

- **Dooley, D. J., Donovan, C. M., Meder, W. P., & Whetzel, S. Z.** (2002). Preferential action of gabapentin and pregabalin at P/Q-type voltage-sensitive calcium channels: inhibition of K+-evoked [3H]-norepinephrine release from rat neocortical slices. Synapse. 250-272

- **DUFAU, M.** (2020). Thesis for the state diploma of doctor of pharmacy: Misuse of Pregabalin in drug addicts.

- **Dussaud, S.** (1998). La méthylènedioxy-3-4-méthamphétamine (ecstasy): revue bibliographique. PhD Thesis .6.

- **Fabritius, Marie.** (2014), Controlled administration study of cannabis and kinetic profiles of cannabinoids in biological fluids. PhD thesis. University of Lausanne,5-6-7.

- **Gandilhon, M, Néfau, T.** (2016). MDMA: an emblematic substance for some young people. *F. Beck, Jeunes et addictions*, 48-50.

- **GHOMARI A, OUSSAHLA R.** (2019). Prevalence of psychoactive substance use among professional drivers in Algeria. Doctor of pharmacy thesis. Abou Bakr Belkaid University. Faculty of medicine. Dr. b. BENZERDJEB - Tlemcen :135.

- **Gothuey. I.** (2014). Cocaine dependence. Etat de lieux .13p.

- **Goyette, M., Charbonneau, R., Plourde, C., & Brochu, S.** (2013). Conjuguer réadaptation et justice : un défi possible. Thinking outside the box: promising practices with adult addicts. Québec : Presses de l'Université Laval, 75-96.

- **Gruffat, X.** (2022). Tramadol (Tramal and generics).

- **Hägg, S., Jönsson, AK and Ahlner, J.** (2020). Current evidence on abuse and misuse of gabapentinoids. Médicament Saf 43, 1235-1254

- **Hickey C, Thomas B.** (2011). Delirium secondary to pregabalin. Gen Hosp Psychiatry Vol 35.

- **Holloway, K. R., Bennett, T. H., & Farrington, D. P.** (2006). The effectiveness of drug treatment programs in reducing criminal behavior: A meta-analysis. Psicothema, 18(3), 620-629.

- **Ives, R., & Ghelani, P.** (2006). Polydrug use (the use of drugs in combination): A brief review. Drugs: Education, Prevention, and Policy, 13, 225-232.

- **Jean-Jacques Coiplet,** Director General of ARS Pays de la Loire/ Santé publique France. January 2019. Bulletin de santé publique. Pays de la Loire.

- **John C.M. Brust.Luis Garcia-Larrea.** (2007). Neurological aspects of addiction. Elsevier Inc.2, 315.

- **KADDOUR M.** (2015). Substance abuse and psychiatric comorbidity. Thesis in medical sciences. Aboubakar Belkaid University. Tlemcen.

- **Kandel, D. B., Yamaguchi, K., & Chen, K.** (1992). Stages of progression in drug involvement from adolescence to adulthood: Further evidence for the gateway theory. Journal of Studies on Alcohol, 53(5), 447-457.

- **LANDIE, A.** (2015). Illicit drugs and overdose,13.

- **Lapostolle, F. and Beaune, S.** (2021) Ecstasy: the second wave... (11) ,103.https://doi.org/10.3166/afmu-2021-0309.

- **Laurent, Karlie** (2014). Addictions. Le cavalier bleu. Esscins. 50p

- **Lebeau. B.** (2002). La drogue. Edition cavalier bleu. Paris,61p.

- **Li, Z., Taylor, C. P., Wéber, M. Piechan, J. Prior, F. Bian, F. Donevan, S.** (2011). Pregabalin is a potent and selective ligand for α2δ-1 and α2δ-2 calcium channel subunits. European Journal of Pharmacology, 667(1), 80-90. https://doi.org/10.1016/j.ejphar.2011.05.054.

- **Marilyn Herie, Tim Godden, Joanne Shenfeld, Colleen Kelly** (2010). Addiction Information Guide.

- **Matthieu Idle, Alexander Bentley, Ruth Arenas Mata, Michel Por,Hugues Soames,Philippa Or, Jane Ecuyers, Hélène Parson,Broche Ng.**(2022) ,Pregabalin and Weed.

- **Matthieu Idle**. (2022). Lyrica and cocaine.
- **MAURY, A.** (1993). Thèse d'exercice. P70
- **Mehdi, Djeribi** (2016). Prise en charge du toxicomane à l'officine,5.
- **Michel Landry and Serge Brochu,** (2012). LE CONCEPT D'ADDICTION : L'ASAM ET LA TRADITION BIOPSYCHOSOCIALE DU QUÉBEC.LE JOURNAL SEMESTRIEL DU GROUPE. (20),1.
- **MIKE VAN, TASSEL. (2017).** Impact of anxiety, depression and hostility as components of psychological distress on the typology of adolescent alcohol and cannabis use by age and gender. PhD thesis. Université du Québec,4.
- **MILDECA** (2021): https://www.drogues.gouv.fr/que-nous-dit-la-science-des-addictions.
- **Miller, W. R., Yahne, C. E., & Tonigan, J. S.** (2003). Motivational interviewing in drug abuse services: a randomized trial. Journal of consulting and clinical psychology, 71(4).
- **Miller, W., & Rollnick, S.** (2006). Motivational interviewing practice. Aider la personne à engager le changement. Dunod. 241 p.
- **NEDJARI M. (2009).** The comorbidity of schizophrenia and drug addiction. A prospective study of one hundred patients. Thesis for a doctorate in medical sciences pp 8-17.
- **Nora D. Volkow, M.D.** (2011). Principles of drug addiction treatment: A research-based guide DIANE Publishing. Third Edition. pp 2-8.
- **PACZESNY, Marine** (2014). Cannabis sativa L.: Botanical and chemical study. Medical properties and regulatory status. Doctoral thesis. UNIVERSITÉ JOSEPH FOURIER, 23.
- **Palma, A., & Pancheri, P.** (2007). Anticonvulsants in anxiety disorder and the role of pregabalin. Giornal Italiano di Psicopatologia,
- **Perry, A. E., Darwin, Z., Godfrey, C., McDougall, C., Lunn, J., Glanville, J., & Coulton, S.** (2009). The effectiveness of interventions for drug-using offenders in the courts, secure establishments and the community: A systematic review. *Substance Use & Misuse*, *44*(3).
- **Richard, D., Senon, J. L., Hautefeuille, M., & Facy, F.** (1998). L'héroïne. Toxibase, Vol 1, 25p.
- **Ries, R.K., D.A. Fiellin, S.C. Miller, & R. Saitz** (2014). **The ASAM Principles of Addiction Medicine** (5th edition), Riverwoods (Ill.), Wolters Kluwer.
- **Sandra, P.** (2000). Héroïnomanie et grossesse : la mère, l'enfant et le personnel soignant. Human Medicine and Pathology, P11

- **Schaad, N. Jean, Clermont. And Drolet, inf.** (2014). Marijuana.p2.
- **Shim, J. H.** (2011). Clinical Application of α 2 -δ Ligand. Hanyang Medical Reviews, 31(2).55.

  https://doi.org/10.7599/hmr.2011.31.2.55.
- **Spigset, O., & Westin, A. A.** (2013). Detection times of pregabalin in urine after illicit use: when should a positive specimen be considered a new intake? Therapeutic Drug Monitoring, Vol 35, 137-140.
- **Sueur, D. C.** (2017). State of the art of research on the therapeutic capacities of "hallucinogenic substances" in the 21-th century. *Psychotropes*, De Boeck Supérieur *24*(3), 125-163.
- T **assone, D. M., Boyce, E., Guyer, J., & Nuzum, D.** (2007). Pregabalin: A novel γaminobutyric acid analogue in the treatment of neuropathic pain, partial-onset seizures, and anxiety disorders. Clinical Therapeutics, Vol 29(1), 26-48.
- **Thérèse Bluet Pajot, M. and Laurent, k.** (2015). Quand mon cerveau devient accro" booklet.
- **TOE B.** (2011). Étude du profil des cas d'abus de drogues et leur prise en charge dans les structures sanitaires de la ville de Ouagadougou (Burkina-Faso). Thèse pharm. UFR/SDS Université Ouagadougou; 88 p.
- **Touami. H & Tayeb. N.** (2018). Effects of psychoactive substances (Alcohol, Drugs and Medicines) on certain biochemical and haematological parameters in drug addicts in the Wilaya of Mostaganem. 62-85
- **TRAORÉ A.** (2004). Étude de lien entre VIH/SIDA et consommation de drogues au Burkina Faso. Thèse méd : UFR/SDS Université Ouagadougou : 98 p.
- **Vandendriessche, F., Lecompte, D., De Hert, M., Hulselmans, J., Mertens, C., Peuskens, J., ... & Wampers, M.** (2010). The role of environmental factors in the development of schizophrenic psychosis. *Neurone*, *15*(10), 8.
- **Vincent, D.** (2017) Cannabis: active ingredient.
- **Yann Botrel.** (2021). Hallucinogens: effects, consequences, drugs. Vol 155
- **Zaccara, G. Gangemi P, Perucca P, Specchio L. (**2011). The adverse event profile of pregabalin: a systematic review and meta-analysis of randomised controlled trials. Epilepsy.Vol52, P 26-36.

List of websites

1. https://hajar-ben.wixsite.com/prof-avec-classe/blank-mj622
2. **Addict aide.** (2022):https://www.addictaide.fr/autres-drogues/les-complications-de-la-mdma
3. **HEROIN** (2022) https://www.docteurclic.com/encyclopedie/heroine.aspx.
4. **EMCDDA**, 2009: https://www.emcdda.europa.eu/polytoxicomanie-.
5. "LYRICA (pregabalin)", High Authority of San t https://www.hsante.fr/jcms/c_538415/fr/lyrica-pregabaline
6. Manulife (2014) https://www.manuvie.ca/particuliers/planifier-et-apprendre/sante-mentale/problemes-de-sante-mentale/toxicomanie.html
7. Neuraxpharm 2021. https://www.neuraxpharm.com/fr/troubles/usagesubstances?
8. https://sante.lefigaro.fr/medicaments/3651330-lyrica-300mg-gelule

MIX
Papier aus verantwortungsvollen Quellen
Paper from responsible sources
FSC® C105338

Printed by Books on Demand GmbH, Norderstedt / Germany